THE HIGHS AND LOWS OF FLYING

by

Harvey F. Mace

Edited by Tom Frisque

Published by

AVIATION USK

602 Front St., Box 97

Usk, WA 99180

ISBN 0-9623080-5-6

CONTENTS

DEDICATION

This book is dedicated to my wife Helen, sons Lan and Kip, and to my fellow charter member fighter pilots of the 362nd Squadron, 357th Fighter Group, who are listed below in alphabetical order:

★ Fletcher A. Adams
Harry R. Ankeny, Jr.
Maurice F. Baker
Henry E. Beal, Jr.
Robert H. Becker
Thomas A. Beemer
Joseph E. Broadhead
Robert D. Brown
Robert W. Brown
Leonard K. Carson
Lynn H. Drollinger
★ Hubert I. Egnes
John B. England
Kenneth E. Hagan
★ Arthur M. Lingo
Thomas K. Norris
Gilbert M. O'Brien
★ Davis T. Perron
John D. Pugh
★ Alvin F. Pyeatt, III
Donald R. Rice
Arval J. Roberson
★ Elmer D. Rydberg
★ Konstantin J. Vogel
Robert D. Wallens
Calvert L. Williams

This book is also dedicated to all the mechanics, other support personnel and replacement pilots who contributed so heavily to the success of our mission. But especially it is dedicated to the seven men listed above, whose names are preceded by a star, who so willingly gave their lives for the cause of freedom.

Harvey Mace
Pudding Creek
April 1994

Keith Ryder's R-4 "Firecracker". Photo courtesy EAA archives.

At the end of the runway at Compton Airport, California, a boyhood dream was about to be fulfilled. It was 1948 and I was sitting in the cramped confines of the cockpit of NX264Y "Jackrabbit" ready to take off. She was a Grieve Class racing plane designed by Keith Rider and built by Dave Elmendorf. She was a twin to "Firecracker", which was flown in the prewar races by famous test pilot Tony LeVier against the likes of Art Chester's "Goon", and Steve Wittman's "Bonzo". At idle the crackling B6S Menasco transmitted every pulse through the longerons pressed against each shoulder. With the short tapered wings rocking to the rhythm of the super charged engine, I pointed the long nose down the runway and shoved the throttle home.

THE EARLY YEARS

Los Angeles, California — the City of Angels. Big bustling Los Angeles, my home town. I was born in the Los Angeles County Hospital on June 10, 1921 and raised in the southwest part of town. Vermont and 110th Street was the center of my universe. In those pre-WWII days, the air was clean and clear most days of the year. You could look past the City Hall, which was the tallest building in town by far, and see the mountains to the north and northeast. By going three or four blocks south of my house, you would find yourself in mostly open country, spotted only here and there with an occasional rural home, farm or small community like Gardena, Torrance, Hawthorne, or Inglewood to name a few. Beyond them were the beach towns from Malibu to Long Beach. With the wonderful weather and the beautiful scenery, the area was a perfect place for flying and sight-seeing.

Most any period of the short history of heavier than air flying machines could deserve the title of "The Golden Age of Flying." But to me the period from when I was born to the beginning of World War II really deserves the title. In those early days there was enough open ground in the Los Angeles basin that airports were everywhere. At one time there was a whole string of airports along the west side of Western Avenue that stretched from Burdettes at about 84th Street to the Gardena Airport (or Western Avenue Airport) near Rosecrans Avenue. As a lad, I was particularly impressed with one airport near Century Boulevard. Its main fence bordering Western Avenue was composed of a string of broken propellers stuck in the ground. Obviously the owner was not current in public relations.

Sharing a fence on the south side of Burdettes was Dycer Airport. It was the busiest airport in the area and would figure prominently in my life to come. About four miles almost due west of Dycer was Mines Field. In later years it would become LAX, the Los Angeles International Airport, but in those days it was still just a field with a tower and several large hangers bordering the north side of Imperial Highway. The activity at this field is also strongly etched in my memory.

With so many airports in the greater Los Angeles basin, it follows that a great deal of flying activity could be observed overhead on most any day, especially after the big factories of Douglas and North American blossomed at Mines Field. The spectrum covered everything — military, civilian, and the exotic. It was a time when lighter than air was still in vogue also. If you were lucky, you could look up on occasion and see the slow majestic flight of one of the big dirigibles like the Akron or Macon.

Seeing breathtaking fighters and bombers like the P-12, P-26, the Grumman fighters, Helldivers and Martins was not uncommon. I've seen the huge Douglas B-19 on final approach to Mines Field from four miles away. Its great size made its movement seem so slow that I wondered what kept it in the air.

Of course the vast majority of flying activity was done by the smaller types of flying machines, and there certainly was a large variety of those. The beautiful year round weather of Southern California was made to order for the people that flew. And those that did fly were regarded as celebrities, even those that were so near penniless they had to get by on homemade crow soup occasionally, as some did around the Western Avenue Airport once in awhile!

Those were depression days and on top of that my brother Jerry, sister Naoma, and I were raised in a very troubled family. I can count at least six separations of my parents, some of them dish throwing separations. For many years we were forced to get by on public assistance of one kind or another. This did nothing for my self esteem, to say the least. I spent my prepuberty years as a self conscious foil for school bullies, suffered academically, and in general just had to scratch for whatever fun I could find. Due to the low income of the family, we frequently moved from one house and neighborhood to another. This forced me to attend two different grammar schools (three times each) and, briefly, one junior high school. The junior high school was Brett Harte where one of the young students occasionally sang at the assemblies — her name was Judy Garland.

Woodcrest was my favorite grammar school, however, and that was where I finally insisted that I was going to stay, even if it did mean walking six miles round trip every day.

Around the time that I reached puberty, some subtle changes began to take place that I wasn't even aware of. I had grown like a weed and had reached my adult height of 5′ 11" at a very early age. Also my brain seemed to shift into a higher gear and I began getting much stronger. Just how I finally became aware of the changes was a little bit interesting. During recess out in the school yard one day, one of the school bullies decided that he was going to pants me right there in front of everybody. Being extremely shy, and ashamed of my homemade shorts that were made out of old bed sheets, I decided to fight back for the first time. All of us on the field, and especially the bully, were very surprised over the ease with which I prevailed. Nobody ever pushed me around again.

It was a pleasant surprise to find that I began to excel in most of the sports as well. By the eighth grade, I was a superior sprinter and beat all comers in the intracity junior high track meet that was held at Washington High School. I could consistently clear 20′ in the broad jump as measured by the gym teacher as well. This was all doing wonders for my self esteem which certainly needed boosting.

In those early days, I was fascinated with guns. Although my father wasn't around much, he still managed to come through for me when there was something I really wanted. So it was that when I found an old single shot 22 rifle for sale for three dollars in a local second hand store, he managed to come up with something out of his collection of stuff to trade for it. I was in hog heaven! Getting ammunition for it could have been a problem except that my good buddy, Bill Appleton, had money for ammo but no permission to own a gun. We made a good combination. We had lots of open country nearby that included a dump with lots of tin cans and other targets. We soon became expert shooters. It was rolling country and the shooting up and down hill taught us the effect of gravity on the long shots. The afternoon winds showed us the effect they could have, also. It was excellent training.

I had developed a knack for working with my hands and solving mechanical problems which came in handy when word of my passion for guns got around. It was surprising how many people came up with old guns to give me. They were throw aways with major parts missing, but they looked like gold to me. With a lot of determination, some scrap metal, a pair of pliers, vise, hammer, hack saw, and a file, I managed to make them work. It was very educational and allowed me to become a fairly decent pistol shooter to boot.

Another close buddy from Woodcrest Grammar school was Walt Wiltfong. He was also a gun nut, and he and his dad were great hunters and skeet shooters. It would serve the country very well during WWII as Walt would turn out many a good aerial gunner as a gunnery instructor during the war. Walt and I occasionally went out in the fields for a little impromptu hunting with BB guns that was more walking and plinking than hunting.

One day we decided to scout the field west of Dycer Airport, actually it was part of Dycer. When a J-2 Cub took off over our heads, Walt made the mistake of doing a little dry tracking with his BB gun. The alert pilot saw him and rightfully became very indignant. For all he knew, we might have been armed with real guns. He whipped the Cub around in a tight three sixty and landed about 50 feet from us. I knew what was coming and told Walt to drop our guns in the tall grass and walk away from them. The pilot was no dummy of course, and proceeded to circle us in search of the "stick" we said that we had pointed. I was quick to see that he had seen through my plan so I casually pulled Walt by the arm and stood right over the guns. After a fruitless search in the tall grass around us, the pilot gave us one more word about playing on the airport and took off. We walked off and came back after dark to pick up the guns.

My mother was extremely tolerant of all my antics probably because of our tough family life. And there were antics! Like the time I decided to set an endurance record for sitting high on a tree limb. No problem, have at it. With my girl friend Elaine hoisting up food, and a can to pee in on a rope, I managed to better 24 hours on that dumb limb with nothing to sit on but a flat board.

Then there was my knife throwing period. Actually, I got pretty good at it. I managed to wear a bare streak in the lawn going back and forth to the target in the process. Once when I had enough throwing for the day, I decided to finish with the hardest throw I could muster just to see how deep I could bury the knife in the wood. The 18" by 30" board I was using was propped up against the plaster wall of our house. The strength of the throw was terrific, unfortunately my aim wasn't. The knife just cleared the top of the target and penetrated the plaster and dry wall of the living room, leaving a neat round hole. It then crashed quite noisily into the metal waste basket sitting along side of the easy chair in which my mother was sitting.

Needless to say she was quite startled. I sheepishly went in and explained what happened expecting at least a finger pointing. But my good old easy going Mother just laughed heartily over being so startled.

Then there was my archery period. Not having learned or remembered too much of the previous episode, I once again had the target against the house wall. This time the target was about shoulder high when I decided to end the day with a long range shot. It missed. It skimmed just over the top of the target and slammed clear through the wall. In the process it knocked a picture down which landed on top of my mother's head, who was again sitting in her easy chair. Fortunately I escaped with another good belly laugh.

Then there was the time I discovered a good way to kill flies. I discovered that by placing a match tail first in my BB gun, the match would explode like a miniature bomb when it was fired towards the wall. I also found that if I hit reasonably close to a sitting fly, the fly was instant history. It was great target practice and almost practical — right up until I shot one at too much of an angle to the wall. At that angle the match didn't explode, it merely ignited and landed right in my mother's lap — and scorched her favorite lavender dress. My fly safaris were put to a quick end.

You couldn't be a true Angeleno without having an avid interest in automobiles either, and I had that in spades. Fortunately the country had yet to become over run with lawyers and law suits and people were still willing to stick their necks out a little bit. A friend of my older brother's taught me to drive his 1928 Star when I was ten years old. Two or three years later an uncle, who admittedly was a bit on the wild side, let me drive his beautiful Chrysler convertible in downtown Los Angeles traffic, under his supervision, of course.

It was at about this time that my brother undertook the job of replacing the clutch in his old Chevy roadster. After starting on the removal of the transmission, he just let it sit for a couple of weeks. I finally asked him why, and he said it was because he couldn't figure out how to get the tranny all the way out. One day I decided to have a look while he was gone. I saw what the problem was immediately and had the tranny out in about twenty minutes. After the repairs were finished, I felt that I had earned a little piece of the action so whenever he was gone, I proceeded to learn all about dirt track racing in the vacant lot next door. It was not only fun, but great experience.

When I reached 15 the car bug was really biting. In my neighborhood wanderings, I had come across a very cherry Model T Ford, stored on blocks, that was for sale for ten dollars. Even in depression days, this was a real steal; but a steal is still no good if you don't have any money. Once again my Dad came to the rescue. In talking to the owner, he found that the owner was a camera bug. Rummaging through his supply of old stuff, Dad came up with a tattered old camera. It apparently had a lens of exceptional quality however, because the car owner was more than eager to swap the car straight across for it. I had my first car and did I ever enjoy it! By scrounging through the vacant lots and alleys, I could often come up with enough returnable bottles to buy a gallon or two of gas. This was during a period when gas was available for ten cents a gallon at the cheaper stations. It didn't take me long to find out that exactly two gallons was all the flashy little "T" Roadster needed to get me to the beaches and back. More than once I coasted into the driveway after an outing at the beach.

The desires of teen-agers don't always follow the practicalities of good reason. It wasn't long until the foot pedal shifting of the "T" Model just didn't measure up to the class of a car with a stick shift and a clutch. I just had to have one. After touring around some of the nearby car lots, I finally found a stick shift beauty that I thought I might be able to talk my way into. It was a 1924 Overland touring car that someone had converted into a pickup. The rear seats had been removed, the back sliced open, and a 2×4" floor installed. For weather proofing, a 2×4" and 1×6" roof had been built over the seat. Only a kid of 15 could have seen any beauty in this car. With Pop supervising the negotiations, we found that the dealer was quite happy to trade straight across for my cherry "T" Model. The Overland may not have looked like much and had wobbly wooden spoke wheels, but it did have a stick shift! Actually, I had seen some possibilities for it. By removing all of the wood, and all of the body work, except for the seat bucket, I wound up with a pretty sporty looking open wheel bug, the first of many hot rods. I would eventually be doing things like putting V8 engines in Model A's, building up hot rod pickups or other similar projects for years to come.

An early photo of a Wiley Post, side by side 2 seat biplane. Much later we would have one of our own!

The frequent activity in the skies overhead was not escaping my notice. It is hard to know just when the seeds of infatuation with flying began to bloom inside me, but bloom they did. I'm sure that our family indulging in one of the few cheap forms of entertainment available on an occasional Sunday afternoon had a lot to do with it. This consisted of parking by the fence at our favorite airport, usually Dycer, and watching all the activity going on. Final approach was right over our heads which gave us a good look at the wide variety of aircraft at the field. There were the many Waco variants, Kinner Fleets, Travelairs, Davis DIW'S, and many others that were the common airplanes of the day. There were also rarer ones like a Bhul Bull Pup, Wiley Post, and Curtiss Jr.

At that time the thought of becoming a pilot seemed to be completely out of reach to me. Pilots seemed like supermen, and were held in such awe by laymen that it seemed an almost unthinkable goal to become one, especially to a kid from a poor family.

It didn't keep me from enjoying flying vicariously, however. The sound of an aircraft engine overhead never failed to make me cast my eyes skyward. I was often rewarded with the sight of the rare and unusual such as when the all green Granville Brother's Q.E.D. flew right over my house, or seeing Howard Hughes flying his new racer. One day I wondered about the Kinner Fleet that flew lazily by inverted time and again. It proved to be a lady pilot who was trying to set a new endurance record for inverted flight.

One very memorable family outing we had was to Grand Central Airport in San Fernando Valley. The occasion was an Armistice Day celebration featuring some infantry units complete with some small artillery and machine gun squads. The Army Air Force was represented with Martin Bombers and a squadron of Boeing P-26 Fighters. When the aircraft arrived, they engaged in a simulated attack on the airfield with the bombers sweeping overhead followed by the fighters making mock strafing runs. During all of this, the ground "defensive" units were firing back with blanks. It was a very colorful and impressive display.

As the aircraft prepared to land, the announcer on the loud speaker system repeated in great detail about the virtues of a smooth, light three point landing and how we were about to see a demonstration of just how it should be done by these well trained P-26 pilots. After such a build up, I was very eager to see the landings and made sure I was in a spot to see every detail of the event. The first fighter came in and skipped and hopped its way down the runway. "Well folks, there are some tricky ground air currents out there sometimes, just watch this next one," and the fighter skipped and hopped its way down the runway. It's not that they were really bad landings, they just weren't smooth, light three point landings. After three or four more skipped their way down the runway, the announcer must have begun to run out of plausible excuses, but finally one touched down with a beautiful feathery three point landing. "There you are folks! There you are! There's a wonderful three point landing!" It all left the indelible impression in my mind that if you were going to be an expert well admired pilot you would have to master the three point landing.

As with most young boys, model airplanes played a part in sparking my interest in flying as well. My brother Jerry had a great touch for building and flying R.O.G.'s. The initials stood for rise off ground. They were little skeletal models that could be made from scrap material and were flown indoors. Calling them R.O.F.'s, for rise off floor, would have been more appropriate.

When I was still quite young, he wasn't above making me the butt of an occasional practical joke. One day he and a buddy talked me into going to the store to buy him a nickel's worth of propeller pitch. They convinced me that it was something that would make his R.O.G. fly better (pitch, aeronautically speaking, is the twist of the prop measured in degrees, not something you can buy). I dutifully trotted down to the local store where the puzzled clerk said they didn't have any and perhaps wondered if they might be missing out by not stocking a hot new item!

It was about 1934 when an airshow was held at Dycer and I was there for every minute of it. I was particularly impressed with the aerobatic display put on by a WWI French Nieuport powered by a rotary engine. This type of engine, where the whole engine revolved with the propeller, had no carburetor as such. They were designed to run wide open all the time. The only control the pilot had over the engine RPM was by way of a kill switch on the control stick. Gliding, taxiing, or any other reduced throttle activity, had to be done by periodically killing the engine. it made for a very distinctive display.

"Wrong Way Corrigan" kept his Curtiss Robin in Dycer's big hangar for a while. I remember peering into the cabin at the huge fuel tank and wondering what he had in mind. This was before he made the flight across the Atlantic.

Feeling so short changed at home finally got to me when I was about fourteen. Being close to Navy country made me aware that they accepted some awfully young guys. I figured that new clothes, and a little steady money wouldn't be hard to take, so I decided to try to enlist. There were only mild objections from my mother, after all, one less body to feed and clothe wouldn't be all bad. Since I was considerably underage, some plausible story would have to be concocted to explain why I had no birth certificate. My mother came up with the fact that a little town in Colorado, called Leadville, had burned to the ground with all records sometime in the past. So, for the Navy's sake that would be where I was born. Armed with the fact that I looked much older than my age because of my size, and a reasonable story to cover my lack of a birth certificate, I went down and started the enlistment procedure. For the early part of it things seemed to go pretty smoothly. But, down the line, someone finally smelled a fish and I was shown the door. It was just as well.

The time finally arrived to leave our eighth grade grammar school and move on to the George Washington High School. Here I was faced with the unbelievable opportunity of choosing most of my own courses. Like many students, I fell victim to the temptation of picking all the goof off courses I could cram into my schedule — brilliant things like Floriculture and Brick Laying. Of course Floriculture wasn't all bad, I was the only boy in the class. George Washington was one of the few high schools to offer an Aeronautics Class, so naturally I chose that too. My strongest memory there was helping to cover the wing of a WWI German Pfalz with it's cable trailing edge. The cable is what gave the trailing edge of the wing that distinctive scalloped look.

I also wanted to become a football player, a quarterback naturally, so I signed up for that too. Things did not go well there. The coach was a huge burly all macho type who didn't take kindly to a future quarterback that didn't even swear and resorted to such pseudo profanity as Dadgummit, and Oh Shoot! It was also obvious that if I was to fully utilize my great running speed, it would have to be done in something besides the huge clodhoppers that were the vogue of the day. I somehow managed to come up with a pair of light weight low shoes with the appropriate cleats. I immediately became the laughing stock of all the players with their clodhoppers.

I could do nothing right for the coach. Because of my speed, I would quite often choose to scramble, even though I had called a pass play. In spite of the good yardage I invariably gained, it still infuriated the coach. "Dammit, if you call a pass play, pass it. I don't care if you lose ten yards!" I found it very difficult to go along with such narrow reasoning. It now pleases me no end to note that if today's quarterback has the ability to scramble, he is held in high esteem. Not too long ago, one player was even pointed out as being a little weird for his choice of high top shoes. Times change.

The following year I switched to Gymnastics, a move I never regretted. I loved the sport which had the added advantage of building up your body instead of damaging it. After a couple of years of Gymnastics, it was flattering to learn that I was known as "shoulders" among some of the girls. After the trauma of my early childhood, I found it flattering to be noticed a little for anything. I've often wondered if one of those girls who might have noticed was the rather tall slim girl that was a terrific swimmer. Her name was Esther Williams.

About the same time I switched to Gymnastics, I realized that if I was ever to amount to anything I would have to forego the goof off courses and change to academics. I made the switch with a math major. Discovering that the Air Services demanded at least two years of college for applicants might have influenced my decision. Fortunately, I took to math like a duck to water, and things rolled along quite smoothly.

I was still fairly young when the flying bug really bit. I began hanging out at Dycer Airport for hours on end just watching the Wacos, Travelairs, Lincoln Sports, Taylor Cubs, and a host of others that were the airplanes of the day. After spending so much time at Dycer, I became a whiz at aircraft identification. I could even identify all of the planes by sound. Even two apparently identical airplanes would have just enough difference in sound that I could tell them apart.

Occasionally I would be puzzled by less common airplanes though. On one occasion, I picked up the sound of a very high revving racing plane approaching. I dashed out from behind the hangar to be sure and get a look at it before it flashed past. While staring in the direction the sound was coming from I waited for it to appear, and waited, and waited. Finally a slow moving low wing job of ordinary appearance showed up and landed on the field. It was hard to believe that this was what was making the sound of a racing plane. Upon investigation, it proved to be an Arrow Sport. It was powered by an ordinary Ford V8 car engine that was equipped with a reduction unit to turn a propeller at low revs while the engine whined away at about 4000 RPM.

Eventually I was able to garner my first short airplane ride in Riley's almost brand new Waco F with a 90 Warner engine. It was an experience never to be forgotten. If my enthusiasm needed any further enforcement, it came when my girl friend, Elaine Keene, got us a ride in one of Nelson's Travelairs through the connection of a relative. With both of us in the front seat we motioned to the pilot to do a little aerobatics. I wasn't able to recognize what he did, but I was hooked!

In 1936 the National Air Races were held at Mines Field. I bicycled down for three of the four days of the event. It was a ten mile round trip but worth every mile of it. Compared to Dycer's airshow, it was really big time. All of aviation's racing big guns were there including Earhart, Wittman, Chester, Ortman, Kling, and of course the Frenchman, Detroyat, who seemed to steal the show. The aerobatic stars did their thing too, of course. I was particularly impressed with Harold Johnson throwing the big Ford tri-motor around. I had the opportunity to meet Harold at an Experimental Aircraft Association gathering at Oshkosh in 1981. It was heartbreaking to see him confined to a wheel chair. He had been a great performer.

Milo Burcham was his usual spectacular self as well as the German pilot flying the Focke-wulf parasol. There were others, too, all were very impressive. The military was there in force and the giant snake designs they performed in trail were something to see. I think the roar of all those big radial engines put the first seeds of desire to be a military pilot in my blood. The racing really inspired me too, but, above all I think the aerobatic flying impressed me the most. It left me with a strong desire, or wish, that someday I would have the opportunity to become an aerobatic pilot myself.

Still, the racing was very spectacular to watch. On at least one day I was able to garner a spot practically under the home pylon. What a thrill to watch those short winged racers make snarling, almost vertical turns around the pylon. On one of Harry Crosby's turns, he did what I judged to be a high speed stall that tucked him out quite visibly. It may have just been turbulence, but in any case he slowed up and flew quite conservatively for the balance of the race. Well, it was all material that a boy's dream could be made of.

As time went on, I became a pretty good friend of Riley's line boy. Riley operated a large FBO operation on the south side of Dycer. I'm almost certain that the line boy's name was Leo Singer, who eventually went on to fly for the California Fish and Game. He was a year or two older than I was and had recently acquired his private license. It turned out to be a great connection. He had a strong ambition to become a rated instructor and I had a strong desire to learn to fly.

Riley operated two Taylor J-2 Cubs that he rented for four dollars an hour. At that time my total income amounted to a nickel a day for lunch money at school. When I went to the school nurse with a tooth ache one day, she just about flipped when she learned that my usual lunch consisted of an ice cream bar! At any rate Leo and I reached an agreement. If I saved my lunch money, and showed up on Saturday with the quarter, he would give me ten minutes of instruction for the quarter. I thought I had died and gone to heaven! After many weeks of quarters and any other change I could scrape up by returning bottles or whatever, I was ready to be given the tail spin segment of the instruction that was required at the time.

Riley wouldn't let his young budding instructor perform that part so he did the task himself. He probably wanted to check up on the progress of Leo's work as well. Apparently we both met with Riley's approval because Leo asked me to save enough money for a half hour's flying time as he was going to solo me the next time out.

It was working out perfectly! I had just turned sixteen, the minimum age for soloing and after four and a half hours of instruction was going to be turned loose! It took about two weeks to gather the money for the required student's permit, and the flying time, but I got it and rushed to the airport. I was met with a very long face and the explanation that the C.A.A. had just passed a new rule requiring eight hours minimum instruction before anyone could solo. What a blow! It was one of the many, many new rules that the C.A.A. (later F.A.A.) would be hitting the general aviation community with in years to come. They seemed to be aiming at hindering rather than furthering the cause of aviation. In any case it was a heartbreaker to me. The prospect of struggling for another three hours of flying time at the rate of ten minutes a week was overwhelming.

On the north side of the field was a row of some 18 or more tee hangars. Those nearest Western Avenue were occupied by what we called fixed base operators, people who have airplanes for rent or instruction. First was Nelson with a couple of Tank Travelairs then Bill who had a couple of Kinner Fleets. On up the row, were various individuals with aircraft like a J-2, Porterfield, Waco Ten, Davis DlW, and the like. Most were private owners but some were would-be operators. Among them was a relative new comer named Elwood Welch, he had a nearly brand new Warner 90 powered Rearwin Sportster. My young instructor suggested that I might try hitting him up for the job of line boy. The prospect didn't seem too bright to me but I decided to give it a try anyway. With hat in hand, I plead my case and lo and behold I was hired. In exchange for coming down to the airport after school, and all day Saturday and Sunday, to keep the airplane cleaned, gassed, and ready to go, I would be rewarded with 15 or 20 minutes of instruction per week in that beautiful Rearwin. Life was looking up once again! The only real down side was having to clean the airplane after the occasional air sick customer! The instructor working for Elwood was R. V. Torrence.

In between times, I was free to stand around and watch the activities on the field, and there were activities! One day one of the Travelairs flared for a landing about ten feet too high, stalled, rolled vertically to the right, caught the wing tip on the ground, bounced squarely off its prop shaft, and settled back down on its nose and wheels. No one was hurt but I was sure the airplane had really suffered. Surprisingly, after straightening out one slightly bent motor mount leg that night, they had it flying the next day. I doubt that a modern spam can would have survived as well.

On another day, one of Dycer's Cubs came in a might too low and caught the inevitable power lines on the approach end of the field with its landing gear. Without the necessary bulk to crash through the wires, it was unceremoniously thrown to the ground well past vertical which smashed the forward part of the cabin and broke the back of the airplane. It did not look good for the occupants. Surprisingly the student was unhurt and the instructor in the front seat escaped with a cut on his forehead.

The time that Bill took a customer up for a sight-seeing ride in one of his Fleets made for a day to remember also. I watched him land and shut the engine off right where the plane stopped in the middle of the field. Then he got out of the plane and stomped back to the hangar with a very grim face-alone. It wasn't sight-seeing the customer had on his mind, it was suicide. When a local housewife got over the initial fright of a horrendous crashing noise, she walked into her kitchen to find the dead body on her floor and a very large hole in her roof! It took Bill quite awhile to recover from that incident, but there could be little doubt that it took the housewife considerably longer!

Airport rail birds, and I admit to being one, are very quick to pick fault with how other pilots are handling their airplanes, often unjustly. I remember watching a Kinner Bird coming in on a high final one day. The pilot was side slipping like mad because his approach was high, and yet he still had so much throttle on that it made the prop spin so fast I couldn't even see it turning. As it turned out, there was a good reason for not seeing the prop, it had fallen off and landed in a vacant lot near Manchester and Normandie! Within a few minutes of the successful forced landing, a gentleman drove up with the prop sticking out of the rumble seat of his roadster. He was so intrigued over all the happenings that he started taking flying lessons himself!

Miles and Atwood Special. Photo Courtesy EAA Archives.

There were several rare airplanes that called the airport home. One was named the Wiley Post and was made by the Straughn Aircraft Company. It was a very lightweight side by side biplane powered by a converted Model A Ford water-cooled engine. The owner sold rides in it for 50 cents a ride. Another rare bird was the Aeromarine Klemm that was powered by a very tiny nine cylinder radial engine of 40 horse power. It too was a very light airplane that was built for modest putting around. Unfortunately the pilot tried to take it beyond its flight limits one day and pulled a wing off, killing both aboard.

I was especially excited another time watching Lee Miles bringing the little Miles and Atwood racer in for a landing. There was something about the special racing planes that was a cut above everything else.

There were other interesting and exciting things from time to time but few, for me, could match the progress of my own training. A day finally arrived when Torrence asked me how many hours I had as I was taxiing in. I responded that if we counted the time to engine shut down, it would be eight hours even. He didn't respond and I taxied on to the hangar and was reaching for the switch to shut down. "Don't shut her off," he said and got out. "Take her out and shoot three landings and bring her back in." I was speechless! Trembling with excitement I did what I was told! It's impossible to describe the joy a raggedy 16 year old can feel during those moments of glory! I taxied back in and was almost in tears with happiness as I shut down and got out of that beautiful Rearwin! I could hardly wait to go home and share the joy with my mother.

The time finally arrived and I hurried into the house and proudly announced: "Mom, I soloed today!" She looked up from her needle work briefly. "That's nice." That's nice? that's nice? The biggest moment of my young life ever and I only get a "That's nice."? Well, you can't always be a hero I guess.

A few days later I circled low over the house several times, jazzed the engine and waggled my wings thinking that might impress somebody, but no, nobody even noticed. Then a couple weeks later Elwood decided to buy a Cub that was located at the Long Beach Airport so we flew down to pick it up. He flew the Rearwin back and I flew the Cub. Our house was right on the way back but I had given up trying to attract any attention and just passed over it at about 600'. When I got home, my mother asked if that was me flying over the house in that little yellow airplane. I was surprised that someone had finally noticed. I answered that yes it was. "Oh, who was flying it?" "I was," baffled a bit by the question. "No,

I mean who was your instructor?" "There wasn't any instructor, I was all alone." Her eyes bugged and her jaw dropped. "You mean, you mean you've been up there flying those things all by yourself!?" At last, a small bit of recognition and the reason my previous solo announcement had only gotten a "That's nice." She hadn't even realized what soloing meant. At last, the family was beginning to realize they had a budding pilot in their midst.

My first solo, a Rearwin Sportster.

In hangar flying with Elwood one day, I mentioned that I had overheard some guys saying that pilots in the south were getting paid to fly. It seems farmers were willing to pay good money to have poison dust spread over their crops by airplane. I asked Elwood what he thought of the idea. His response was immediate and emphatic. "Don't even think about it! Not unless you're ready for an early death. Those Cropdusters are getting killed right and left." And that is where the subject was dropped. Still, the idea of flying and getting well paid for it didn't seem to be that bad an idea to me.

Population pressure was already making its presence felt as airports along Western Avenue disappeared at a rapid pace. Western Avenue and Dycer Airports were the last hold outs and now Dycer's days were numbered. The ground where I had soloed, and where the likes of Curtiss Pushers, Lincoln Pages, and Buhl Pups had made so many takeoffs and landings, would soon be covered with houses and shopping centers. All too soon we were forced to relocate to Mines Field. It was a really big league airport compared to Dycer. Several of the racing planes of the day were based there. The Miss Los Angeles was hangared next to us and flown fairly regularly on Sundays by Marion McKeen.

The last of the Gee Bee Supersports was just beyond that. Actually it was built up from the remains of both the Rl and R2 and was being flown by Cecil Allen. I sure was awe struck at the size of the Gee Bee. It was an immense beast. I wasn't there when Allen took it up for his first flight in it. The rail birds said that it was a touch and go thing and that he made four passes at the field before he got it back down. Further down the line was the little Popjoy powered Light Airplane Developers F-15. Boy was it small, especially after looking at the big G.B.

Another interesting goings-on at our hangar was a set of batman wings being built for a sky diver named "Tommy" Thompson. Batman jumps were the going rage at air shows in those days. The usual accouterments for the jumper was a battened web of sail cloth between his arms and body and between his legs. Thompson had a better idea. Instead of the arm webs, he was having a set of miniature wings built of about ten feet span. These were made to fold forward, brought together by hand holds at arms length so he could back out of the door of one of the larger planes. After clearing the plane, the idea was to stretch out his arms and wings and glide around like a bird until the time came to salvo the wings with a rip cord. He then planned to finish the descent by parachute. Or so the plan went. Before the wings could be finished, a couple of other batmen had convinced themselves that they could actually safely land on the ground with their web wings. Needless to say, they were dead wrong.

As a result of those accidents, the C.A.A. banned any further batman jumps. This put an end to any official attempt for Thompson to try

The Brown B-2, "Miss Los Angeles", left, and the Light Airplane Developer's F-15 "San Diego Flaggship" were two of the more interesting residents at Mines Field in those days. Photos courtesy EAA archives.

his wings. Of course, with so much effort having gone into the project, Thompson would not be denied. He took his finished wings out to the desert to give them an unofficial try. After backing out of the plane, he stretched out the wings which promptly folded together behind him. So went another good idea. He saved himself by quickly pulling the wing's rip cord and parachuting the rest of the way down in a much more conventional manner.

Mines Field had many of the same kind of planes that Dycer had, but also some classier stuff like newer Fairchilds, Stinson Tri-motors, and the like. There was a beautiful all silver Ryan ST flown by a pretty young gal all dressed up in an all white jumper and white helmet. They made a striking picture. There was also a Waco F that someone had installed a menasco engine on that apparently changed the center of gravity because it ended up on its nose more than once. I remember watching a Fairchild 24 groundloop which collapsed the right gear and scraped the wing tip. It sure made me wince.

We had only been at Mines Field a short time when the C.A.A. came up with another of its beauties. They decided that student pilots needed a solo license to fly solo. Of course you had to solo first before you could try for your solo license in order for you to fly solo. What a dumb idea. Well, I had to go for it. The check consisted of the Inspector sitting safely in the control tower and watching me make three takeoffs and landings through the window. I made three of my finest landings and passed. Searching for a compliment I immodestly asked Elwood how the inspector liked my landings. "Oh! the inspector never watched, we were too busy talking". It sure seems difficult to be a hero!

Fortunately, someone in the Bureau soon saw the stupidness of the solo license and it was abandoned. Anyway, that leaves me as one of the few pilots that has ever had a solo license.

Finally the day came when I could actually fly how or where I pleased. No practicing this, that, or the other unless I wanted to. One of the things I had wondered about was just how little throttle you could get by with in a really steep turn. Not yet being fully cognizant of all the potential hazards of flying, I looked for the answer over the open fields north of Mines Field at about 400 feet. Fortunately, for my young life, I've always had a very good feel for flying and stopped the experiment just short of disaster.

Aviation seems to breed more than its share of hams, or is it just that we hams are drawn to aviation? I remember a fellow back at Dycer who would show up on Sundays dressed in the finest Roscoe Turner tradition—breeches, leggings, military style jacket, inevitable white scarf, helmet appropriately turned up over his ears, goggles, and a seat pack 'chute tucked up on his backside for walking. When enough spectator cars had lined the fence, he would slowly parade up and down in front of the line, pausing occasionally to shade his eyes and look up into the wild blue yonder. When he felt that he had impressed the crowd enough, he would then strut over and climb into his J-2 Cub!

One of the operators at Mines Field was similarly afflicted, only his passion was aerobatics. He had a reasonably aerobatic Fleet. He would periodically step out of his office on week ends and check the size of the spectator crowd. When he felt that it had reached a size worthy of his performance, he would take off and climb to just north of the field and regale the crowd with his aerobatic routine. Both he and the crowd were rewarded one way or another, so you can't pick too much fault with his ego.

It wasn't too long after we moved to Mines Field that it became obvious that there were more than enough FBO's on the field, so Elwood and a fellow who operated two Kinner Fleets decided to merge. Since my duties as line boy had now doubled, the conglomerate's executives decided to up my pay to ten dollars a week and a half hour of flying . This was wealth beyond my fondest dreams. In fact it was embarrassing and made me feel a little unworthy. True, I was relegated to flying the Cub, though I did get a half hour ride in one of the Fleets. I was surprised that the Kinner engine made the Fleet vibrate to the point that there were two horizons. I didn't mind being limited to the Cub though, certainly I enjoyed the Cadillac like ride of the Rearwin, and the Fleet was aerobatic but I did love that little Cub.

About this time I discovered that one of my fellow gymnasts, Tom McGerty, was also a pilot and had about 37 hours of flying time. Evidently neither of us had talked much about flying at school. Before long we learned that some recruiters from Canada were in town looking for pilots to join the Eagle Squadron. We talked about it a while and decided it would be a whale of a way to get some great flying, and get paid doing it. I'm afraid national causes had little to do with our decision to go down town and apply. Neither had we given much thought to the danger that war might have to offer.

When we went down to apply, I was quite disappointed to find that they required an applicant to have 35 flying hours and I only had 17. Since I was too honest and naive to lie, I was rejected. What in the world could the difference of 18 hours have made in view of the extensive training that would be necessary to make a fighter pilot? Tom was accepted though, and became a member of the Eagle Squadron. Unfortunately, and very sadly, Tom became a victim of the Luftwaffe and died in his Spitfire under the guns of an ME-109 over the English Channel. It seemed like such a short time after he joined the group that I learned of his death. After the war, a movie

United States Army Air Corps P-39 Airacobras. Bell Aircraft photo.

was made about the exploits of the Eagle Squadrons. At the beginning of the movie, pictures are shown of those that had fallen and Tom can be seen quite clearly, donning his helmet and goggles. Unfortunately he would not be the last of my friends to die over Europe.

Although the older fighter planes, like the P-26 always made my heart beat faster, the newer aircraft coming out of the factories prior to WWII were even more impressive. I remember watching a news reel that featured a new fighter plane manufactured by the Bell Aircraft Co. It was the P-39 Airacobra, billed as a new 400 mile an hour fighter. As it flashed across the screen, it certainly looked every bit that fast. As a youngster in the theater, it seemed totally out of reason to even think of ever being able to fly such a beauty.

Eventually the summer of 1940 came around and it was time for High School graduation, a big event in anyone's young life. By this time, I had been casting fond eyes at the possibility of eventually becoming a fighter pilot for the good old U. S. But the good old U.S. required a minimum of two years college training before they would even consider accepting an applicant. At any rate, Fate seemed to have other plans for me. In order for my mother and me to get off welfare, I would have to find a job.

While scouting around for job possibilities one day, I found myself driving south on Sepulveda Blvd. As I approached the end of the runway at Mines Field, which at that time was just beyond a chain link fence near the road, an airplane sitting in the run up area made me slide my car to a screeching halt. There sat one of the sleekest looking fighter planes I had ever seen. The idling in-line engine was making the kind of crackling noises that inline engines often make. But the noise quickly shifted into a smooth purr as the pilot throttled up, moved on to the runway and took off. The fighter very rapidly disappeared into the wild blue yonder. I was impressed! What kind of luck would it ever take for someone to find themselves flying such an airplane? The airplane proved to be North American Aviation's new fighter, the NA-73X and was no doubt being flown by Vance Breese. It would soon become one of the most renowned fighters of the war, the P-51 Mustang! I had just seen history in the making!

The academic pressures of the last year of school, along with cars, girls, and beaches had forced me to give up my treasured position as line boy. But the urge to fly would never leave me. It would just have to wait for this period of readjustment to be put behind me. I finally landed a job at Douglas Aircraft just across Imperial from good old Mines Field again. No, it wasn't as test pilot, or even line boy on the flight line, it was in the maintenance department. But what the hey, it paid real good money. Actually, as it worked out, I really enjoyed the job. It was phenomenally good experience for a kid just out of high school. In fact it's too bad all kids couldn't go through the experience. I learned how to spread gravel with a dump truck, repair plumbing, work a jack hammer, do carpentry, and all manner of other things. In a few months I even had my own helper.

Of course in doing my job, I frequently had to walk around the airplanes being built. They were Dauntless Dive Bombers for the Navy, designated SBDs. I had no way of knowing it then but these airplanes and I would cross paths in a most interesting way in the future, a most interesting way, indeed.

The good money meant that I could have renewed my flying interest but before I could get to it Pearl Harbor came along and private flying was banned from all coastal areas. I had to content myself with buying some really great cars and spending many happy hours at the local roller skating rink. It was there I met a girl named Helen Johnson. I took her home and she very cleverly left her skates in my car to make sure I would be back. It worked. I lost all interest in any other girls.

GEARING FOR WAR

Pearl Harbor came along and my first taste of "war" was the great Los Angeles air raid! I was working the night shift and saw the whole thing. Search lights and antiaircraft fire flashed everywhere! I stood out in the open, the better to see it all, until the rattle of shrapnel on the tin roof made me decide being under a roof was a better idea. Excitement bordered on panic in many areas. What a fiasco! Damage was wide spread all over town, though not severe. All of it from the antiaircraft guns! No trace of enemy aircraft was ever found or proven. This would not be the last of the antiaircraft fire I would see, however.

Soon, news trickled down that the Air Corps had dropped the college requirements for cadets, providing you passed certain written tests. With my academic background, I felt I had a good chance of making it and immediately made plans to sign up.

On the way down to the recruiters, I picked up a hitchhiker about my age and in the ensuing conversation he learned of my intentions. He laughed and said, "You may as well turn around and go home, I tried the same thing several weeks ago. Out of a group of about 30 to 35, only six or seven made it past the first hurdle. Then survivors were grouped to another 30 or 40 and again only about six or seven survived the second hurdle. This went on until about the 5th hurdle and they dumped me. Give it up, you'll never make it. I couldn't believe it was that tough and forged ahead anyway. It was quite a surprise when it worked out exactly as the hitchhiker had said, with but one exception; I made it!

After surviving all the hurdles and finally being accepted, it was a little discouraging to find that the pipeline was full and we would have to wait a while before we could actually begin training. However we were given the choice of staying home until we were called or joining the Air Corps as Unassigned Privates at 21 bucks a month. Always the thinker, I cleverly reasoned that those that opted to join would be the first to be called, especially since those that didn't would be no burden to the service. Wrong. I guess it was kind of like picking the short line at the bank.

After choosing to go ahead and join, I was sent to Fort McArthur, a comfy base, and only 18 miles from home. There were about eight of us that made the choice and since the people at the Fort didn't know what to do with us, we were immediately given a three day pass, in fact several passes in a row. I was feeling quite smug over my choice.

There were some days when we had to stay on the base. So, to kill time, a few of us went to the gym. One day, a big burly guy came into the gym in boxers garb, along with another guy who appeared to be his manager. After the boxer had warmed up a while, doing what boxers do to warm up, the manager came over to where we were and said his fighter needed to do some sparring, would any of us care to help him out? Nobody else raised their hand, so feeling charitable, I volunteered.

There were several tactical errors that I made, not the least of which was stepping into the ring in the first place. The second came right after the bell to start the first round. I shuffled up to the guy in the best fighter stance I could muster, and hit him. I quickly learned some valuable lessons. First, even well trained gymnastic muscles are not what's needed for the type of strength and stamina required for boxing. Second, if you're a 5'11", 160 pound gymnast, you shouldn't get in the ring with the 230 pound heavy weight champion of the Army—which this guy turned out to be. Thirdly, and perhaps most importantly, don't ever try to beat a man at his own game. After two and a half rounds, my arms were so fatigued I couldn't raise my hands above my waist, I just had to cry uncle. I didn't feel too badly though, even after being antagonized the champion of the Army wasn't able to knock me down, much less knock me out.

Soon the Fort's brass realized they had some good manual laborers sitting around doing nothing. As a result we found ourselves removing good lawn from some of the local cemeteries and using it for camouflage around gun emplacements in the Palos Verdes hills. Then there was the day they decided to test the huge pieces of artillery guarding the Fort. Unfortunately, I missed it, but I did get a good look at the aftermath. The guns were located near the rear of the Fort. This meant they had to fire over the top of some of the buildings. I can't be sure what damage might have been done to the target, but surely it didn't match the damage to the Fort. The concussion flattened the theater and took the roof off of several barracks!

For reasons unknown to me they soon decided to move us to another camp. Maybe we laughed too much over their gun testing, or maybe they found themselves short of barracks! In any case it was like moving from Heaven to Hell. We found ourselves in the very hot tents of Camp Maar somewhere in the middle of the California desert and in mid summer. It was not a happy camp. The principal occupants were the Army's most dangerous criminals being guarded and served by others who were awaiting disciplinary action, or being disciplined for one thing or another. This included the camp Commanding Officer. Talk about feeding lambs to the wolves.

We were immediately put to work doing guard duty and KP, lots of KP. As baby faced future officers, we were looked upon as fair game. After going on one of their frequent drunks, the C.O. and his drinking buddy from the mess hall loved to go around to our guard posts. While the C.O. would try to distract us, his buddy would sneak around and try to disarm us which would have meant instant disciplinary action. We may have been young, but we weren't dummies. All they accomplished was to turn us all into the most trigger happy guards the camp had ever seen. We fired warning shots at anything that moved suspiciously and didn't halt on command.

Part of the guard duty was taking prisoners out on work duties. One day they assigned two of the meanest looking murderers to my high school buddy, Jack Kincaid. Jack was about our smallest, youngest looking future cadet. When the prisoner saw him he said, "Why Hell, I'll break him in two and feed him to the coyotes!" Jack never blinked, just shoved a shell into the chamber of his rifle, pointed it squarely between the eyes of his would be tormentor and quietly said, "You just go right ahead.," Jack had just earned himself a whole lot of respect.

$21 a month "Private, Air Corps, Unassigned".

Thankfully, we were eventually assigned to Camp Hahn near Riverside, a complete opposite to our desert outpost. It was definitely vintage "California laid back," a very easy going atmosphere. For lack of something better to do with us, we were once again assigned to guard duty. Since all of the regular guard stations were already adequately taken care of, they elected to have us guard an area that consisted of mostly empty blocks of the camp. Our new superiors equipped us with rifles and assigned some of us two to a block. I lucked out by being assigned as assistant to the O.D. (Officer of the Day).

On the first night, when things began to be a little boring, the O.D. suggested that we ought to run out and check the guards. It sounded good to me, so we jumped into a staff car and the O.D. proceeded to charge off towards the guarded area. As tactfully as I could, I suggested that it would be prudent to slow down and approach the guards at a much more cautious pace. My suggestion fell on deaf ears. As soon as we reached the first station, four guards appeared in the headlights with rifles leveled straight at the windshield. The startled O. D. screeched to a stop to the words of "Dismount and be recognized!". They were spoken with great authority. Not realizing the seriousness with which we had been forced to take our guard duty, the O.D. leaned around the windshield and said:

"Aw Hell, you guys know who we are." Bang! A warning shot whistled inches over the top of the vehicle. It no doubt was the first warning shot ever fired in Camp Hahn! With that the O.D. and I quickly dismounted and dashed in front of the headlights to be recognized. On the way back to the office I hastily explained why our guards were so touchy. Fortunately he took it in good humor. Eventually, we managed to become a little more civilized, although I think that some of the scars from our Camp Maar experience would linger forever.

Eventually the day came when we could leave behind those months of "Air Corps Private, Unassigned." An opening finally came up in the training pipeline and we were sent to Montgomery, Alabama to begin the first phase of training, called Preflight. There was a flying field on the premises so we could at least see a little flying activity from time to time. A pay raise to $75.00 a month and a change to dress uniforms that were identical to regular officer's, except for the insignias, was really a boost to our morale. We occasionally were even mistakenly saluted by newer enlisted men. We had come a long way from Camp Maar.

There was another side to Preflight, however. It was here that we learned to give up a whole lot of our personal freedoms, like the freedom to dress sloppily for instance. The upper classmen saw to that! You soon learned to polish the brass parts of your uniform to a high sheen with a "blitz cloth", and keep them that way. Shoes had to be shined to perfection. You never stepped out of your barracks without first polishing your shoes — ever. Even if you just stepped out briefly and went back in for something you forgot, you polished them again or else! Also our beds had to be made up drum tight. If an upper classman walked by and dropped a coin on your bed, it had better bounce! Great care had to be taken with how you tucked your shirt in your pants. The line of your shirt had to exactly line up with your belt buckle and the fly of your pants.

Under the stern and watchful eyes of the upper classmen, virtually all of our activity would have to be done with square corners. No cutting diagonally across open spaces from one place to another. You walked at attention and made square corners. This would make eating for the next month a difficult chore indeed. We had to march into the mess hall in single file, find our seats and wait until given permission to sit down. Then, in due time, we were given the O.K. to start eating. All of this had to be done with square corners, at attention, and with eyes straight ahead. The food was wonderful, but for the next month we would never see it any better than the way we could see out of the corner of our eyes. Naturally, since the food had to be scooped or speared under these conditions, and raised to our mouths with a square corner action, not all of it was going to get to our mouths.

Our days were completely filled. We had long hours of marching while singing the appropriate songs, and we engaged in more types of physical activities than one could imagine; endurance runs, push ups, obstacle courses, etc. Fortunately, my athletic background helped me zip through this phase in good order. In addition there seemed to be a constant barrage of physicals and mental tests. If all that weren't enough, there were the classes. Lots of classes. A good deal of this activity was conducted under the watchful eyes of the harassing upper classmen. Not unexpectedly there were quite a few of our classmates that would not be able to survive the pressures and tests. Unfortunately the washout rate all through Preflight would be quite high.

One of the tasks we had to undertake was one that I eagerly looked forward to and that was going out to the firing range. We were to run through the qualification course and try for a medal firing the rifle. In view of all my practice shooting, I expected to quality for an expert medal in a breeze. What I didn't count on was the breeze that can blow in Alabama at that time of year. On the day we went out, the wind was gusting to at least 25 to 30 MPH giving a wind chill factor of well below zero, and well below what I was dressed for. As a result of all the shivering and shaking I'm not sure I confined all of my bullets to the shooting range, much less the target.

The only time I ever saw a little forgiveness for a break in discipline was when we were in the field marching one day. Word had rapidly spread that a twin engine Curtis AT-9 trainer was preparing to land with its landing gear stuck in the up position. Everyone, including the upperclassmen, watched with bated breath as the pilot brought the craft in for a smooth belly landing. we all broke out with cheers and applause. In the months ahead I would become so hardened to such things that I would watch such an event without a flicker of emotion.

It was tough to put up with the harassment of the upperclassmen but we could temper our displeasure with the knowledge that in due time we would become "the Devil." There was a certain amount of fun attached to it all anyway. When the time came to become upperclassmen, we had to go through the rigors of voting in the class officers. I became the Cadet Sergeant, which made things a little easier. Not that anything could make going through Preflight easy, but at least we were finally on the downhill side..

During what little time I had to myself, I found my thoughts going back to Helen more and more. I was missing her dearly. I began flooding her with marriage proposals to the point that I'm sure she must have thought the only way to stop them was to say yes. Her treasured response finally came and I was ecstatic. Since I was about due to report to Primary Flight Training at Decatur, Alabama she agreed to meet me there. We were optimistically hoping that we could overcome the Air Corps's inherent distaste for married Cadets.

When I arrived at Decatur, I was pleasantly surprised to find that we would be training in Stearman PT-17s. They were not all that different from the prewar biplanes I was so familiar with. Primary training was handled by civilians under contract to the government. However there was still Brass around to oversee things. Brass that needed to be convinced I should be given permission to marry, and a pass to do it in. As a matter of course, passes were virtually non-existent in Primary, much less marriages. It was only after much haggling that they finally relented and gave me a six hour pass. Six hours!

It was somewhat amazing to find that six other cadets came to Decatur from the Los Angeles area and several were even from my High School. Walt Uhl was one, who was also one of the few married cadets. He and his wife, Peggy, kindly agreed to stand as witnesses for our hastily organized wedding. It does seem to be a small world.

Poor Helen, she not only would have to put up with a makeshift wedding, but she would get gypped out of any semblance of a honey-moon. On top of that, she would have to find herself a job just to be able to afford to stay around. Fortunately she has always been very talented and resourceful. She would be my flying buddy for many years to come.

Winter was coming on in Alabama. It was beginning to give new meaning to "The Four Seasons" for this Southern California boy. I didn't know people lived in places that got that cold!

After finally getting settled in at Decatur, we were assigned, in groups of five, to an instructor. Luckily I drew a real nice guy named Knapke. I say lucky because five students meant the instructor had to spend five hours a day in the very cold open cockpit of a Stearman with no heat. It would have tested the resolve of an Eskimo! After a day's work, even easy going Knapke would occasionally end up a might testy.

Unlike the British, the U.S. didn't like Cadets having previous experience, actually it was frowned upon. They didn't want the instructors having to undo any previous bad habits a civilian trained pilot might have locked in his brain. If it was going to be done, it was going to be done the military way! The Stearman was quite a bit more airplane than the Rearwin but I was able to adapt to it quite quickly. Being an honest airplane, it was easy to fly.

What was not easy was getting the 220 Continental engine on the Stearman fired up in the bitter cold of an Alabama winter. Many mornings would reveal icicles hanging off the trailing edges of the wing. The cold would make the oil in the engine about the consistency of molasses. The engine was equipped with an inertial starter which consisted of a flywheel that was brought up to very high revolutions by a hand crank. It was then engaged to the engine which would then turn over fast enough to start usually. If it was a warm day, this could be accomplished by one reasonably able-bodied man. But on these cold mornings, it took as many cadet hands on the crank as the crank handle would accommodate. They had to push as hard as they could push for about a minute before the crank would even start moving. It was a back breaker, especially when a balky engine refused to start even after two or three tries. Even young healthy bodies could be pushed only so far.

As a last resort the slingshot was brought out. This consisted of a very large "Y" made of very large bungee chord with a leather cuff fitted to the bottom of the Y that fit over the tip of the prop blade. With one cadet holding the prop in the proper position, several other cadets on each of the other two legs of the Y would then run off, stretching the chord to it's maximum. At this time, the cadet on the prop would release it and the prop would snap around and hopefully start the engine. It usually took several tries before the comforting sound of the engine could be heard. I don't recall any of our cadets having a weight problem.

Which brings to mind the food at this station. The Air Force was always obligated to finance the food supplies for the cadets. By some quirk of inefficiency, the civilian operators were also required to finance the food. As a result there was double money for the mess hall and we cadets were the beneficiaries. We were fed as if it was a luxury hotel. Each meal featured great food and a choice of several types of fresh baked biscuits, and waiters to serve it. The table cloths, napkins, and multiple choice of quality silverware was almost too much. But we learned to bear up under it. It was fortunate that we had plenty of physical activity. Anyway, enjoy, it was the last station where we would ever have such luxury.

I had no trouble with the early phases of instruction and was soon soloing the Stearman. This did not mean the end of our having dual instruction, there were other things ahead like aerobatics. I was really looking forward to getting into that. A slow roll would be the toughest maneuver to learn. A lot of my time in bed was spent mentally going through the necessary control movements to complete a slow roll. The more simple moves like Lazy 8's and loops would not be quite the challenge.

The only method of communication in the Stearman was by hand signals from the instructor in the front seat, or verbally through a Gosport which was one way only. All the student could do was look and listen. In due time we were up on my first aerobatic dual ride. Naturally the more simple stuff came first. Things went smoothly with those. Fortunately, I've never had any tendency to become air sick, so that helped too. It wasn't long until Knapke got around to demonstrating a slow roll. Then came the words through the Gosport I had been waiting to hear. "Now you try one." As carefully as I could, I put my mentally rehearsed moves to work and luckily produced a reasonably acceptable slow roll. There was a short pause from the front cockpit and then Knapke turned around to look me in the eye and said: "All right, just how much flying time did you have before you came here?"

I couldn't answer of course, I just smiled smugly to myself. We went on with the rest of the usual routines, punctuated here and there with practice forced landings. It seemed like he pulled those forced landings on me more often than necessary, but I didn't mind, they were no problem. When we returned to base to land, he frequently challenged me to land on a line even with some marker on the edge of the field without using the throttle. I don't recall ever missing by more than would be acceptable. Having mastered the basics quite well, I began spending less time on them during my solo flights and more and more on aerobatics. If you work at practicing most anything long enough, you're bound to get pretty good at it.

Eventually Knapke felt that I had progressed enough that we wound up spending quite a bit of time just playing around. I remember my eyes being opened wide on one flight when he did a fairly low roll over his girl's house. I'm sure it must have been the tenseness and unfamiliarity with the low altitude that caused him to un-characteristicly dish out of the roll. Fortunately we started the roll high enough to allow him to recover. Impromptu low altitude rolls would kill many pilots in the years to come.

Not many pilots are ever thrilled about having to endure a check ride. Some weather them pretty well and some don't. One of my high school buddies, Vernon Jordan, was a very good pilot but was one of those that just tightened up too much on a check ride. He was washed out unfortunately, thus depriving the Air Force of a potentially very good fighter pilot. He went on to become a very highly decorated infantry officer while fighting in the "Battle of the Bulge". He was awarded almost every medal an officer can win including two Purple Hearts. He was not the only cadet to wash out, of course. Quite a large percentage of our original class would not survive the rigors of the classes and training. Many a heart broken cadet would find himself transferred to bomber or navigator school.

With Primary Training coming to an end, we came to what had become a tradition at this base. The last hour of flying would consist of the cadet becoming the instructor and our instructor becoming the hapless student! We all looked forward to this with maniacal glee. We did our best to ask the impossible from our "student" and then made great use of the Gosport to emphasize his total ineptness and lack of effort!

Following this, was an informal party where hair was let down and anything could be discussed in a friendly atmosphere. Knapke came over to me and put his arm around my shoulder and said: "Mace, you're probably as good a student as I've ever had, but why in Hell did you always miss the forced landings?" I was dumb founded! "What do you mean!? I always picked the biggest and best plowed field around and never missed!" "Plowed field! Plowed field!? I said pick the best sod field! That's grass!" "Well, heck, I thought a sod field meant plowed!" What did I know, nobody from my area ever used the word sod for grass. The only farm term I had ever heard was hearing a farmer referred to as a "sod buster." That only invoked visions of a plowed field in my mind. "Well cripes amighty! If I ever get another student from California, the first thing I'm going to do is teach him what a sod field is!" No wonder he seemed to pull more forced landings on me than I thought were necessary!

Our stay at Decatur was ending and the move to Basic Training back at Montgomery had to be undertaken. Poor Helen, the move for me was routine, but she had to terminate her job, move out of her room, find her way down to Montgomery, locate a place to stay, and find another job. I'm always amazed at her patience and resourcefulness.

Helen's Viewpoint

After deciding to accept Harvey's proposal, I was in no mood for delay. It seemed the more time we could spend together before he went overseas, the better it would be. A fancy ceremony was the least of my priorities under the circumstances. It was wartime and people had to accept limitations. Our wedding on that Thanksgiving Day at the quaint little parsonage was very meaningful to me and I felt as confident of our future happiness as any wartime bride might feel. Harvey and I had talked of the prospects of marriage and found we shared many values and interests. I loved him dearly and felt that I could be happy with him in that lifetime partnership.

Being a cadet's wife was an eventful and very happy time for me. Any problems were handled without difficulty and time with Harvey was delightful. It was immediately obvious that the living costs were such that some extra income was in order. Checking with other cadet wives made the chances of finding a job seem grim. Only two that I knew of had found jobs, one with Kress's and one with a grocery store where her duties were mainly taking telephone orders. Sounds strange in these days of modern computerized super markets. There was certainly no office work available in this small town. The approaching Christmas season may have saved my day because with persistence I found a job at the only other variety store in Decatur.

Having other cadet wives for get-togethers was certainly a plus. I had found a board and room arrangement with Mrs. Robb, a warm southern lady who was writing a book. She allowed me to have a little gathering of some cadet wives in her home. The wives seemed to be from far flung states and of varying backgrounds. Peggy Uhl, who stood up with us at our wedding, was from Los Angeles as we were. The youngest of the wives was the daughter of a doctor. One beauty, who had spent years studying ballet, now appeared to be a little chubby. I suspected she was pregnant. I was being very guarded in this area since I wanted my husband around if possible when the great occasion of a baby's arrival occurred. We all knew that our men were scheduled to go overseas as soon as training permitted.

Cadet passes were really rare in Primary so any time Harvey and I had together was treasured. Most pilots like to talk about flying and this suited my interest. I had been intrigued with aviation since Lindbergh's historic flight to Paris. Later when I was in the 6th Grade and living in Chickasha, Oklahoma, Wiley Post was scheduled to be at the airport one Sunday afternoon. My brother and two cousins and I walked over six miles to see him on that cold day. The sight of Wiley Post climbing into that biplane with his black eye patch contrasting sharply with his white scarf fluttering in the breeze made a mental picture that remains with me to this day.

My first airplane ride occurred the next year. Our family had moved back to Marlow, my home town. When I read in the Marlow Review that a barnstormer would be giving rides that weekend for a penny a pound, excitement really reigned. How could such an airplane ride possibly be managed? If my father were still alive, he would have treated the whole family but now funds were extremely short. My mother finally gave me permission to spend my treasured Indian-head penny collection for this thrill of a lifetime. It proved an even greater experience than I had imagined. There was the nearly unbelievable take-off into the blue, then the circling of Marlow. As I looked out and saw all that detail below of the school buildings, train station, water tower, tiny vehicles, etc. I was entranced.

That bit of flying, modest as it was, helped make listening to Harvey's flying experiences even more enjoyable to me. I never ran out of questions about how one avoided stalling, or other fine points of flying and Harvey never tired of explaining the mechanics of flight. Much of our visiting was by nightly telephone calls where we shared our activities of the day. In Primary, it was possible to see the Stearmans doing aerobatics during walks on the edge of town. I described them to Harvey and learned about spins and Immelmans from him. During the few treasured passes Harvey had, there was much to share and we were the happiest newly-weds you can imagine.

Any regrets over leaving Decatur, where I had nothing but fond memories, (well almost nothing, my memories of the cold were certainly not fond ones) were tempered by the fact that we were moving on to a bigger and better airplane, the Vultee BT-13, an all metal low wing monoplane powered by a Pratt and Whitney 450 HP engine.

Once again there were all the preliminaries to go through. There were classes, briefings, and physicals before we would even get a look at the airplane. One of the classes was for learning how to use the E6B computer, a little hand held device designed to aid flying cross countries. A major difference at this base was the fact that it was all military, including all but one of the instructors.

Eventually we were assigned to our flying instructors and once again it was five cadets to each instructor. Mine was named Lt. Martin and when he called out "Mace, Martin, May" during the first roll call, he stopped short and said: "My gosh, what a coincidence! when I went through Basic, it was also Mace, Martin, and May." It was quite a coincidence since the name Mace is not all that common.

A trio of BT-13s.

Unlike Primary, we would not keep the same instructor all the way through Basic. I'm not sure why they insisted on changing things around so much. It not only was disruptive to a smooth line of instruction, but it also prevented any kind of rapport. The instructors were 2nd Lieutenants, for the most part, with a minimum of experience. This would also be a disadvantage since Basic would involve more complex flying than Primary, including night, instrument, and formation flying. Later on I had one hour of dual with the lone civilian instructor who seemed to be a little older than the rest of them. Since it was only for one hour, he chose to do most of the flying and just fooled around. I have to admit that he was the smoothest pilot I had ever ridden with. Just riding with him did a lot towards helping me smooth out my own flying.

Eventually we were taken out to be introduced to our new mounts, an impressive looking airplane. Compared to the Stearman it was huge, twice as powerful, and even had flaps. It also had a full canopy which was very welcome after the cold at Decatur. We had to go through the usual cockpit familiarization routine. With two trim tab controls, mixture control, fuel tank selector, two position propeller, and a fuel pressure gauge that were all new to us, we certainly needed the coaching.

Actually the BT was a very straight forward flying machine and after getting over our initial apprehensions, things began to go quite well. We were soon put through the basics once more. Though nothing was too different than what we had done at Primary, it was necessary to get familiar with the larger, faster airplane. There is no doubt that war made the training less than what would have been desirable, but on the whole it did the job.

One glaring exception was the instrument flight training program. Too many of the instructors lacked confidence in this area themselves and consequently often resorted to yelling at their students instead of imparting a calmer, more productive, step by step approach. I suppose that we students lacked the proper motivation, too. It was hard to imagine that we would ever be required to do any blind flying.

Ordinarily we flew the BT-13 from the front seat but for instrument training, the student flew from the rear seat which was equipped with a hood that completely blocked out any view. I particularly recall one occasion when I was scheduled for an hour of instruments with a temporary fill-in instructor. I started out feeling quite satisfied with myself because no matter what maneuver I did I seemed to be holding the ball of the turn and bank practically dead center with surprising ease. Yet, that maniacal perfectionist in the front seat started complaining immediately and was yelling quite impatiently by the end of the hour. Since he was in the front seat he did the landing and taxiing which gave me time to just look around.

Suddenly I noticed while taxiing back that not even "S" turning would make the ball move, it was stuck! Another feature the BT had over the Stearman was an intercom. I quickly picked up the "mike" and reported my find to the instructor. I expected a quick vindication. All I got was: "Yeah, yeah, sure!" in a very derogatory tone. Well no matter, I could show him as soon as we got back to the ramp. The thought had hardly passed through my mind when we hit a large bump in the taxi way that shook the ball loose! My morale sunk to a new low. Fortunately the instructor must have decided that since he didn't have to fly with me anymore, he would just ignore the whole thing. He simply walked off and that was the last I ever saw of him.

In addition to the poor quality of the instrument training, we were required to put in lots of time in the Link trainers. These were ground bound poor facsimiles of airplanes designed to teach instrument flying and procedures inside the safety of a building. As for any similarity between "flying" these and a real airplane, there wasn't any! The only thing they were good for was teaching us how to bracket a beam on a radio range. This we practiced by the hour. Why, I can't imagine, there certainly would be no opportunity to use one in the real world of the Air Force.

The BT was a reasonably good aerobatic airplane and I spent as much time as I could doing all the aerobatic maneuvers I could dream up, just as I had in Primary. Even with what seemed like an unnecessary amount of instructor changes, progress went along at a reasonable pace. I particularly enjoyed formation flying. One of the instructors, who was especially hot on the subject got me going. Before long, he decided to show me how he could touch another airplane. The BT had quite a prominent pitot tube that stuck out and down from the leading edge of the wing out near the tip. He jockeyed up very close to the other BT we were flying with and after three or four tries managed to lightly touch the other BT's wing tip with the end of the pitot tube. By this time I had acquired a reasonable amount of cocky confidence about my flying. This is a virtue that some people might find objectionable, but in fact, it is a very desirable one for a fighter pilot to have. At any rate, I felt that I could perform the "stunt" more smoothly than he did. (I hate to use that word "stunt" but I can't find a more suitable synonym). At the first opportunity, I gave it a try and feel that I succeeded.

As a matter of fact, many of the things we pilots engaged in while training might seem to have been nothing more than frivolous games to the layman. Indeed, some of them were games, but they were not frivolous. Anything that sharpens a pilot's wits and skills is highly worthwhile and might someday save his life, or that of someone else.

Even though I have poked fun at pilots who are a little overly concerned about the image they present, I have to plead guilty to a certain amount of the same concern. One thing that always struck me as being real "cool" was the use of a white scarf by the prominent old time pilots. It did have its practical purposes back in the open cockpit days. Anything that helped keep you warm was definitely O.K. The white scarf was so indispensable that it finally became a pilot's badge, much like the stethoscope is to a doctor, or a wad of chewing tobacco (yuck!) in the cheek of a ball player.

By the time I reached Basic Training, I felt entitled to wear one and began to look at it as my "lucky scarf." But then there was the day that I was just flaring the BT for a three point landing with the canopy back when the scarf suddenly blew up and completely covered my face. While frantically clawing my face clear there was a moment of doubt about the scarf being "lucky".

The BT was quite a heavy beast with a fairly slim symmetrical airfoil. If it was handled unwisely at low altitude and low airspeed, it could result in disaster for the pilot. More than one light plane pilot who acquired these aircraft after the war would find this out the hard way. One day I was heading in to land after an hour of dual with one of my almost regular instructors when he picked up the mike and said: "Mace, you seem to have a very good feel for flying so let me show you an alternative to flaps or side-slipping for losing altitude. The idea is to lower your gliding speed by raising the nose, like this, until it is just above a stall. See, it lets the airplane mush down like an elevator. O.K. take over and try it on a high final to the field." I felt extremely flattered that he was willing to trust me with the whole ball of wax on such a tricky maneuver that close to the ground. Fortunately I didn't let him down.

Among the other things we were expected to accomplish while in Basic Training was to master the art of night flying. I have always had mixed emotions about night flying. It's an interesting challenge on the one hand, but at the same time flying is a matter of seeing and being seen. Staggering around in the dark with a lot other planes in the area is just not conducive for much piece of mind.

Nevertheless, it had to be faced and the time had come to face it. If we students approached it with some trepidation, it had to be nothing compared to what the instructors must have felt. They had to sit in the back seat of the BT where it was extremely blind in broad daylight, much less on a dark night—and sit there with a student flying from the front seat who had less than 150 hours of flying time total.

There were several satellite fields in the vicinity of the home base and this is where the checking out would be done. The reluctance the instructors had for this phase was amply demonstrated by the fact that if the student managed to take off, find his way to the satellite field, and land without hitting something or busting the airplane, he was immediately proclaimed "qualified to solo" by the instructor. He then quickly got out of the airplane and sent the student on his way. Fortunately most everyone survived so I guess it had to be rated a qualified success. However, not every cadet would survive all of the trials and tribulations of Basic Training and there would be a noticeable number of wash outs. Not as many as in Primary, but the number was substantial none the less.

With Basic Training successfully behind me, it was time for still another move. This time it was on to Dothan, Alabama. Poor Helen, she had to once again pull up stakes, sever all ties behind her and resettle in Dothan. What heart!

Helen's Viewpoint

As our train headed toward Montgomery and Basic Training for our husbands, Molly (one of the cadet wives) and I became better friends. We talked of sharing a room in Montgomery until she could get more funds from her folks and until I could find a job. We found a room in a pleasant neighborhood and it worked out well as Molly was good company.

I was experiencing the usual rebuffs hunting for a job. After all, who wanted to hire someone who was expected to be in the community only a short time. It occurred to me that the local business school might have a referral and surprisingly they did. A young lawyer, just starting out, needed a secretary but wages were very low. My thinking was that a low paying job was better than no job. I went in for an interview, took a quick test in dictation, and was hired. It turned out great, I wanted to maintain those office skills. Molly had been on the ball and found separate rooms for us in a home near the bus line including kitchen privileges for breakfast and lunch. I dashed off to work each day.

The best thing about the Montgomery base was its policy that allowed cadet wives to have dinner with their husbands at the base. It was cafeteria style but the food was good and there was a great recreation room so it gave us a chance to visit each night. What a boon! Actually there weren't many married cadets so it probably had little impact on their costs. I took the bus out to the base each day after work and our time together resembled regular married life more than it did in Primary.

Harvey was getting Link trainer lessons now and was thrilled with the BT. He somehow managed to get me in the Link trainer one night and the Sergeant ran a little test to show me how it worked. I promptly spun out but that was not surprising considering I'd never touched the controls of an airplane. It did give me a better understanding for further airplane talk. Passes were much more frequent in Basic and we managed to do a few things like going to a movie and roller skating. They were happy times for us.

Next, we were sent to Napier Field in Dothan for Advanced Training in North American AT-6s. However, we cadets would have to endure more tests, physicals and classes before we could get a good look at it. It was worth the wait though, that AT-6 was a gorgeous looking airplane. In addition to all of the features of the BT, the AT had a constant speed prop, hydraulically controlled flaps and a retractable landing gear that we would have to master. Unfortunately, my start in AT-6s would be pretty rocky. If I thought instructor changes in Basic had been bad, I hadn't seen anything yet. For the first five hours, I had a different instructor each hour and since the instructors knew that it was just a temporary thing, their only interest was in playing around. As a result by the time everyone else was soloing the AT I had hardly touched the controls. When I was finally assigned to a permanent instructor, I'm sure he must have been taken aback by my ineptitude.

Fortunately there were two things in my favor. I was a quick learner, and the instructor was a mediocre pilot at best, so my apparent ineptitude was quickly overcome. He was an RAF pilot on detached service to learn the American way of flying. His main purpose for being in the "Colonies" seemed to have gone astray, however, as I often saw him leaving the base lounging luxuriously in a big American convertible with a pretty American girl doing the driving. Well, who could blame him?! After getting a good hold on the basics of flying the AT, I spent a good part of my solo hours doing aerobatics. It was a real pleasure to fly the AT and with it's ample horsepower, it was a good aerobatic airplane.

Instrument training in Advanced was pretty much the same as we had in Basic, poor. We were again subjected to flying the beam in the Link trainer and suffering through still another refresher course on the same old E6B computer. But I sure enjoyed flying the good old AT-6. It was especially pleasing to fly close formation with another AT. With the gear tucked up and it's almost bulldog look, it didn't take much imagination to think of it as a fighter airplane, and a fighter airplane is where my heart belonged.

Besides my rocky beginning in Advanced, there was one other period of concern that I suffered through and that was the night cross country. As mentioned before, I was never all that thrilled about night flying anyway. Part of the curriculum called for a solo night cross country, so it had to be done. I really can't recall the length of the trip we were supposed to make but it was a triangular course that I believe was something on the order of 50 miles a leg. After much careful plotting and planning, two other towns were picked as the other two points of the triangle and the check points were carefully noted. Variation on the map and deviation of the chosen aircraft's compass were taken into consideration along with the winds aloft, and off I went.

It was an uninviting pitch black night but at least the weather was clear. The first leg went well, check points like major highways and small villages came up where they were supposed to and soon my first destination appeared. The second leg went as smoothly and I was soon heading for home and none too soon as far as I was concerned. The first check point came up right on schedule and the second one was only a little late. But where was the third one? Perhaps the wind had come up just a little stronger. I decided my only alternative was to press on. After the time allotted to reach home base had nearly gone by, there was no home base in sight. Horror of all horrors, I was really lost! No layman can fully appreciate the helpless feeling a situation like this can give a pilot. However, being a Los Angeles driver, I knew better than to ever give up.

By scanning the horizon, I was able to detect a faint glow to my right and one to my left. They had to be towns of some sort. My only hope was to pick one and head for it — and hope like the devil that they had an airport. With little delay, I picked the one on my right and headed for it. As I got closer there was something a little familiar, what — yes! It was home base and I was practically right on time! A power failure coming to someone sitting in an electric chair couldn't have been more welcome! I landed, taxied in and parked with a feeling of relief that is hard to describe. I said nothing to anyone, my night cross-country was behind me.

The time had come for my RAF instructor to expose me to the rigors of aerobatic instruction. He had no way of knowing that I was already well versed in the intricate art of throwing an AT-6 around. I should have been, I certainly had practiced enough. I even managed to innovate a little. I discovered that a fairly simple loop could be made into a much flashier maneuver by tossing in a snap roll at the top. For many years I thought I had originated this move since I had never seen nor heard of any one doing it before me. But alas, while reading about the exploits of some old timer not too long ago, the article mentioned the maneuver as being in his repertoire. It's mighty tough to be first at anything in aviation.

We climbed to an appropriate altitude over the approved aerobatic area and cleared the area below us. My RAF instructor then proceeded to demonstrate all of the usual maneuvers and managed to fall out of every one of them, even the relatively simple loop. he followed each one of them with the explanation: "Well, that wasn't very good but you get the idea." He finally turned it over to me and I began my effort with the loop, just as he had done. I wasn't able to resist a little one upmanship so I threw in my snap roll on top, and went directly on to the rest of the maneuvers, ending with a slow roll. He picked up the mike and asked: "How in the Hell did you do that slow roll without falling out of it?" We spent part of the hour with me trying to teach him how to do a slow roll.

Just before quitting for the day, I did another snap roll and on the recovery things just didn't feel right with the rudder. A few jabs on the rudder pedals confirmed that I had lost all control over the rudder. I picked up the mike and reported the problem to my instructor expecting him to take full charge of the situation. He gave the pedals a few jabs and said: "By golly you're right. Well.... uh, well go ahead and take it back and land it." That was a surprise, but it was all right with me. The airport at Napier was a large square grass field (grass not sod!) with a wind sock right in the middle. This made it very easy to land directly into the wind. Landing an AT-6 directly into the wind is not a bad idea even if you have rudder control. After making sure I had the traffic pattern to myself, I lined up directly into the wind and very carefully landed. I had mastered the art of a smooth three point landing a long time ago, so after touch down it was a fairly easy matter to control any desire the AT had to swap ends with light taps on the brakes. At that time I had very little interest in the mechanical side of airplanes so I never took the time to find out what caused the failure. There have been many occasions since then that I wished I had.

We were nearing the end of Advanced Training and the time had come for us to receive the gunnery phase of the curriculum. Eglin Field, Florida, is where this activity took place so it meant a brief move for us cadets. Naturally you never moved far from calisthenics and marching and it was on just such an occasion that I had my second good look at a P-51. Apparently the pilot couldn't resist showing off to a bunch of green cadets assembled on the ramp. He made a sizzling buzz job over our heads at a speed that had to be close to 500 miles an hour. I was impressed. That airplane was screaming like a jet! Of course at that time I couldn't have known what a jet was. The surprise of the unexpected buzz job caused a lot of nervous giggling on the ramp.

It wasn't much later that we were treated to a vision on the opposite end of the spectrum. An L-4 (J-3 Cub) came "roaring" down the runway tail high and at full speed which was maybe 85 MPH. When he reached a point near us, with his wheels about two feet off of the runway, the pilot pulled the L-4 up into a nice loop and quickly — no, make that leisurely — disappeared in the distance. It was quite a contrast.

The AT-6s at Eglin were equipped with a single .30 caliber machine gun in the nose.

After the usual classes, briefings, et al, we were finally able to fly an actual armed aircraft. We then did our best to hit targets on the ground and in the air. Ground gunnery is much the simpler of the two, at least without flak coming back up at you.You just need to stabilize your approach path with the pip on the target and you are bound to get numerous hits on the target. Hitting an airborne target sleeve, which is towed behind an airplane, is another matter entirely. You can't just drive up behind it and shoot as you might at an enemy aircraft — tow pilots would get pretty cranky if you did. This means that every shot has to be a deflection shot with constantly changing angles and lead. As near as I can recall a score of 15 to 20% was pretty acceptable, and 25 to 30% got you an expert shooters medal. In any case, I was able to qualify for an Expert Medal in aerial gunnery and I felt pretty good about that.

One of the last things we did back at Napier Field was to go through the test of a mock dog fight with one of the check pilots of the base. The outcome of the mock dog fight might figure heavily in the decision of just what our next assignment would be. Although we would be graduating from a single engine advanced school, there would be no sure bet that we wouldn't be moved on to some multi-engine assignment. The Air Force's decision's were not always totally logical.

The start of the dog fight consisted of the two opponents approaching each other at the same level and to the right of each other. As soon as they were exactly opposite the fight was to begin. It would end as soon as the "guns" of one could be brought to bear on the other. It invariably ended with the check pilot winning. There were three other pilots scheduled ahead of me which suited me just fine. It gave me the chance to go out and observe just what tactics the check pilot employed. It was a smart move as it allowed me to see just what helped him always win.

In the first place he cheated by breaking into the attack before he was supposed to. This tended to fluster the student who was a little too tentative anyway. He also used a lot of vertical reverses that allowed the student to overshoot and end up in front of his guns. When it came to my turn, I was ready. I had learned to be defensively offensive a long time ago. I just watched his wing tip and the instant it started down, I broke into him. I immediately throttled back and dropped 30 degrees of flaps. He soon was under my "guns" and was cold meat, according to the rules of battle anyway.

The washout rate in Advanced was down to a trickle and most of us were due to graduate. Discussions of just where we might go were frequent. I wanted fighters with all my heart, but there was absolutely no way of knowing. Graduation was coming up, along with our commissions to Second Lieutenant. The ceremony was going to be very elaborate and I was asked to give an aerobatic performance over the festivities. Of course! When did a pilot ever pass up a chance to show off?

Helen's Viewpoint

Dothan, Alabama, where Harvey was to take Advanced Training, was an even smaller town than Decatur. It made me wonder if my luck at getting a job would hold out just one more time. Because of my of having already worked for a lawyer in this state, my hopes were high. But after nearly a week of making the rounds with no luck, I began to wonder.

One day after lunch my landlady asked if I had tried the attorneys, Lee & Lee. I replied that I had earlier in the week but had been turned down by the girl in the office. She suggested that it might be a good idea to try again. It didn't seem too likely a chance, but for lack of other possibilities, I decided to try. As I found out later, the landlady had privy to information that I didn't have. It seems their regular "girl friday" had joined the Army and the replacement girl did not work out and had just been let go. I went back to their office and this time I talked to Mr. Alto V. Lee. It turned out that he had known the attorney that I worked for in Montgomery back in their college days. He gave me a little test for dictation and hired me that afternoon. The Lees turned out to be great employers. I learned a lot on the job and it proved to be good background for later jobs in the big city of Los Angeles.

My landlady also referred me to a friend of hers who took in boarders for dinners only. It was a homey arrangement eating with her family and a few cadet wives. We gave her our ration coupons and she managed to prepare wonderful home cooked meals enhanced by her home grown chickens and small garden.

The cadet wives that I met in Dothan were a different group than I had known before. Earlier friends, like Molly, had gone with their husbands to other bases. Molly had been having morning nausea before we left and she suspected she was pregnant. We kept in touch by correspondence for quite a while afterwards.

The lunchroom of the Dothan Drug Store was a popular meeting place for us wives. They made a delicious chicken salad sandwich and it made a great way for me to catch up on the latest news. When the talk got around to going to Kelley Springs, that really peaked my interest. It was a tradition in Dothan for most offices to close mid afternoon on Wednesdays, at least during warmer weather. The only air-conditioning in our office was provided by a big ceiling fan! There was bus service to Kelley Springs and as it worked out, I was able to go there often. With the help of these outings, our little group became quite close even though we knew our time together would be coming to an end before long.

There were certainly things one could find to complain about in the South. In winter, the coal dust was quite annoying. One generally could ignore the dirty air but I was reminded of it each night as I washed my silk stockings and saw the clear water turn black. Also cockroaches seemed to be everywhere. Back in Montgomery, I took a small can of juice to work each day for break time. One time I hadn't finished my pineapple juice when I decided to go down the hall to the ladies room. When I returned to my juice to take a sip, I came awfully close to getting a cockroach in my mouth. It was many a moon before I could take a sip of pineapple juice again. However, the gracious southern people I met made up for any other shortcomings.

A proud young Second Lieutenant H.F. Mace right after graduation.

The summery weather was a boon for Harvey's overnight passes, which he was getting more frequently than before. The cool evenings after a warm day were ideal for our walks. It was a time when we could share the current happenings and wonder about where his next assignment would be.

Graduation time arrived on a beautiful sunny day in May. A small group of us cadet wives stood together as we watched our husbands march out in perfect form for the ceremony. I was proud as a peacock when those coveted wings were pinned on Harvey.

MOVING UP

Advanced Training was all over and we had our travel orders. I was ordered to report to Hamilton Field, California. It was fabulous to think of being back in California, but there still was no way of getting a hint of what type of unit I would be assigned to, or what type of airplane I would be flying.

Once again Helen had to break camp and we headed home. The Air Force had allowed me enough travel time that we were able to go to Los Angeles and pick up the trusty old '33 Ford twodoor that I had sold to Helen when I left for what was then the Army Air Corps. In my eagerness to see what the future had to offer, we motored up to Hamilton Field a day early and drove into the sight of row after row of C-47s! What had the Air Force done to me!? I was heart sick! There was no way that I was going to report in a day early to be stuck with those. Like a whipped dog I turned around and drove out. The Air Force would just have to wait one more day to rub my face in the dirt.

I spent a very miserable afternoon and night. Even being back in California could not lift my spirits. I'm sure Helen would have loved to have spent the time somewhere other than with a grousing husband. Morning came and there was no way of avoiding the inevitable. I had to report. Helen drove me back to the base and I checked in.

"Oh yes Lieutenant, you are assigned to the 328th. They are located at the last hanger down at the south end of the field. " It was a long walk past one row of miserable looking C-47s after another. I couldn't have been feeling lower as I neared the last hangar. As I got closer, I could see the tip of a wing past the corner of the hangar. It was much too small and low to be a C-47, what... wait! Now the small rounded shape of the tail was visible. "That's a P-39!" I almost shouted it out loud! Two more came into view! A sign over the door confirmed it, the 328th was a P-39 outfit!

My faith in the Air Force had been restored and I was walking on cloud nine! As eager as I was, I would have to wait a few days to even sit in a P-39, however. The checking out process wasn't done at Hamilton, it was carried out at the Oakland Municipal Airport. The usual preliminaries had to be gone through like ground school, a thorough study of the operations manual, and a thorough cockpit check. The cockpit check included having to be able to reach every switch, handle, and instrument on request while blind folded. The Air Force didn't want your head in the cockpit groping for some item when your attention was needed outside. At least for the moment, I was able to stand very close and look at a real fighter plane, The Bell P-39 Airacobra. It was the same P-39 aircraft I had drooled over while watching the news reel in the Balboa Theater. It didn't seem to be all that long ago.

The P-39 was a unique airplane. It was small as fighter planes go and the engine was mounted behind the cockpit with a drive shaft passing under the seat to a gear box in the nose. This allowed a 37mm cannon to be installed that fired through the hub of the propeller. In addition there were two .50 caliber machine guns mounted above the cannon which were geared to fire through the propeller. A pair of .30 caliber. machine guns were also mounted in each wing outside the propeller arc. Later models had the four .30's replaced with two .50 caliber guns. Either version offered considerable fire power. I just stood there and found it hard to believe that I might actu-

Hamilton Field; P-39Qs of the 362nd Fighter Squadron, Hayward, California Summer 1943. Photo courtesy of M. Olmsted.

ally get to fly this wonderful piece of machinery.

Since it was apparent that it would be a while before I was settled in any one place, we decided that it would be best if Helen went to stay with her Aunt Gussie in the Monterey Bay area. For the first time since we were married, I was making enough money that poor Helen didn't have to rush around looking for work. She certainly deserved her break by the ocean. Over in Oakland I was finally going to enjoy an experience most pilots can only dream of, checking out in an honest to goodness fighter plane. There were eight of us greenhorns that were due for the experience and a long drink of water, who really looked too tall to fly a P-39, had been saddled with the task of checking us out.

A couple of the other Advanced Training schools had one or two war weary P-40s and a few lucky cadets were able to check out in them before they graduated. Numbered among the eight of us was one of those lucky ex-cadets. When you have nothing hotter than an AT-6 in your log book, anyone with ten hours of fighter plane time is a real veteran pilot. Our esteemed tutor must have thought so too and placed him first in line. I'm not sure that there was any special reason, but he lined me up as number two. That was certainly fine with me.

The P-39 was a hot little fighter and all hot airplanes need lots of respect, particularly by inexperienced pilots. Our check out pilot went to great lengths to caution us about this, even to the point of telling us that the P-39 would not glide. "Do not try to glide this airplane, if you get it too slow it will stall and kill you. This airplane will not glide." That seemed strange to me, but he should know. It was fairly late in the day when the magic moment arrived and number one taxied out and took off. This was my first look at the take-off of the P-39 and it looked magnificent. I could hardly wait for my turn.

Eventually the allotted time for number one's flight was over and he bent the airplane around on a fairly far out base leg and turned on to his final approach to the runway. But my

Maintenance crewman Sgt. Scott leans on the wing of "102" at Hayward. Merle Olmstead Photo.

goodness he was whistling. Did he feel confident enough to be buzzing the runway? No, his wheels and flaps were down, still, he flashed over the runway much too fast to land. He came around again — and again, each time he managed to get a tiny bit slower and lower. It was now very obvious that he was not completely in charge of the situation. On the fourth pass, he was halfway down the runway when he finally succeeded in getting low enough to force the P-39 onto the runway. he was still much too fast for the length of the runway that remained. All he could do was lock up the brakes and hope. Smoke poured from the skidding main wheel tires, then sparks flew from metal parts of the wheels after the tires had worn away. Fortunately the P-39 came to a stop just short of the end of the runway, but flying was definitely over for the day.

To impress the pilot with the fact that he had goofed, the powers that be commissioned him to help the mechanics fix the airplane that night. That put a heavy onus on my shoulders. Now I had all night to sweat over the prospects of the next morning. If a very experienced fighter pilot was having trouble flying the P-39 what in the world was I facing? Well I had asked for it, I would just have to get up in the morning and face whatever that little beast had to offer. It did seem to be a hot and sweaty night. The next morning dawned bright and clear, my trial by fire would not be delayed.

At 5'11", and 160 pounds, I pretty well filled up the cockpit of the P-39 but it was still a very comfortable place to sit. In fact the airplane fit me like a glove, in spite of being designed around a 5'8" pilot, as has been reported. The entrance was provided by doors, very much like those on a car (perhaps not too surprisingly, they were built by the Hudson Motor Car Co. under sub contract!). They were rather unusual for an airplane but it made getting in and out very easy. Visibility was superb and the tricycle landing gear made ground handling a snap.

All the studying, testing, and coaching were finally over and I was ready to fire the beast up! Judgment Day had arrived! The engine instantly responded and I found myself sitting amongst sounds and vibrations that I had never heard nor felt before. This was definitely going to be different. The instrument panel shook so much that it was hard to read the instruments. In taxiing to the run up area, I found the airplane amazingly responsive and easy to handle. That tricycle landing gear was taking all the sweat out of the ground handling. As soon as the pre-takeoff checks were over, the tower cleared me for take-off and I moved on to the runway and shoved the throttle home. This was it!

The first thing I noticed was that things smoothed out beautifully with the increase in RPM. The next thing was the phenomenal acceleration that full throttle produced. Most satisfying! Again that tricycle landing gear kept me tracking down the center of the runway, making this the easiest thing I had ever flown.

The whole take-off was as easy as it gets — and so satisfying!

All of my apprehensions had not been put to rest however. The memories of the previous day were still fresh in my mind, and the handling of this plane was going to take some getting used to. It was so responsive that it gave the feeling of instability. As a result, I limited my turns to very gentle ones. I kind of had the feeling that if I banked it over too much, it might just want to keep on rolling over. it wouldn't, of course, but that feeling, along with all of the cautions our check out pilot had hammered into us, dictated that I proceed with great caution. There would be no hot dogging on this flight! Otherwise the flight was going just great. The only hint of a disappointment came after settling down in level cruise configuration. The airspeed indicator showed only 235 MPH instead of the 400 MPH that the news reel had promised. However, by throwing in the effects of density at the altitude I was flying, the actual ground speed would have been much higher — a lot faster than anything I had ever flown before! So I wasn't really disappointed.

When my allotted time was nearly over, I approached the field to land. With the memory of those smoking tires yet to fade, I admit to plenty of tenseness over the prospect of landing. But in spite of my nervousness, I have to confess that it was very ego inflating to be calling the tower for clearance to land from a fighter plane. Very satisfying indeed! After getting clearance, I throttled back, advanced the prop pitch, and entered the pattern. On the down wind leg I dropped the gear and lowered 20° of flaps. Everything was in the green so I throttled back a little more and gently made the last two turns onto final approach where I dropped full flaps. In spite of the check out pilot's admonition that the P-39 would not glide, it was pretty obvious from the way the ground was rushing by that trying to land this thing with power on was surely not the way to go. It just wasn't going to work that way. I throttled back some more — and some more. Finally the throttle was all the way back and P-39 was gliding. Of course the thing would glide, and glide very well. I knew then who really should have spent the night helping the mechanics. A gentle flair over the numbers was shortly followed by the easy squeak of the main gear tires on the runway. The nose wheel touched soon after, and there was all kinds of runway left in front of me. I quietly went around to the other guys and gave them my findings. There was no further trouble with the check out flights.

My next couple flights still found me being very tentative, I was flying it with a great amount of respect. Before long, while I was still flying like a sissy, a P-39 came zooming by me about 500′ to my right. He rolled over on his back and went straight down. I was shocked, my first thought was: "That fool, that fool! He's going to kill himself!" He didn't of course, and pulled out below me to go merrily on his way. I wasn't quite ready to roll over and follow him down, but it did show me that I was being far too timid with the P-39. I was able to build up my confidence very quickly after that.

Within a few days I was sent on to Tonopah, Nevada for further training in the P-39. There's not much in my memory banks about Tonopah, but then, there's not much in Tonopah either — even today. A few of the flying memories are vivid, however. The runway was at least a mile long and we were located near one end. This meant a long taxi either to or from a flight. Given the mid-summer temperatures and the very marginal radiator for the liquid cooled engine while taxiing, it was necessary to hurry to the end of the runway before the engine boiled over. That would have been simple enough except that the brakes on the P-39 were a multiple disc type and had to be babied because they over also heated very easily. As a result we didn't always get to the end of the runway before the coolant was too hot for takeoff. It was not uncommon to have

Two more photos of P-39s undergoing maintenance at Tonopah Nevada. Photos courtesy Merle Olmsted.

to shut down and cool off before we could proceed.

The flying we did at Tonopah consisted of solo familiarization flights for the most part. Occasionally one of the other pilots and I would get together for some impromptu formation flying. On one such flight, we were well to the west of the field when the other pilot's engine suddenly quit at about 4,000′ over the desert. I followed him down and watched as he made a successful belly landing amongst the sage brush.

I dashed back to the field and asked the tower for an expedited clearance to land so I could give directions to the ground rescue people. The guy in the tower sounded a bit on the bored side when he replied: "Well, you will just have to wait your turn, we have a P-39 on down wind with a jammed gear and he will have to belly in first." I circled the field and watched my second belly landing in about five minutes!

The accident rate was very high at Tonopah. While standing in front of the pilot's shack one day, I was watching the flying activity when I noticed a P-39 on a long bomber type approach. It was a typical approach made by less experienced pilots. A hot shot in another P-39 dove down in front of him and peeled up in the usual 360 degree overhead pattern made by more experienced pilots. He was almost around the full circle when he became aware that he was cutting in front of the other P-39. He gave what appeared to be an irritated jerk on the stick to peel out of the way. The P-39 does not take kindly to this kind of treatment and it snapped into a one turn spin, creating a large crater in the floor of the desert. It was tough to see this terrible accident, but even tougher to believe that I would one day become so hardened to such things that I could view an accident with the same detachment usually reserved for watching the shadow of a cloud slipping over the top of a hill.

As expected, my stay at Tonopah was fairly short, but by the time I had to leave I was very much at home in the cockpit of the P-39. I was learning to love the little beast in spite of its short fuse.

I was transferred to the 357th Fighter Group. At this time the three fighter squadrons of the group were scattered over Northern California. The 362nd Fighter Squadron was stationed at Hayward, the 363rd at Oroville and the 364th at Marysville. The whole group had been stationed at Tonopah having left just prior to my arrival. It was a fighter group still in the building stage and I was assigned to the 362 Fighter Squadron at Hayward. The C.O. was Major Hubert I. Egnes. His flight leaders and operations officer were very experienced pilots and he was in the process of taking in green pilots to fill in the balance of the squadron. He was sorting through these new additions and weeding out the ones he didn't care for. I believe that I was the last one to be accepted. It would assure me of being one of the tail end charleys for some time. Major Egnes was a heck of a nice guy. The only negative was the fact that he rejected my good buddy Bill Martin over some relatively minor goof. Bill went on to be a very good P-38 pilot.

Since this assignment seemed to be a fairly stable one, Helen found a keen apartment in Hayward and moved up from La Selva beach. War time meant scarce living quarters and the "keen" apartment was the converted office rooms of a warehouse by the railroad tracks. But then, we were newlyweds and any place that afforded some privacy was just fine with us. The next few months proved to be some of the most enjoyable of our lives.

The weather and the countryside were so beautiful that we bought a motorcycle to increase the fun of our outings. We soon became acquainted with another couple riders and we had many enjoyable rides together. As luck would have it they just happened to own the local riding stables as well. This was a pleasant surprise as Helen and I really loved horses too. Who could blame me for feeling smug during this period? Here I was, a happy newlywed 22 year old pilot being very well paid to fly a terrific fighter plane in a fighter squadron with a great bunch of guys — and getting to do it all in one of the most attractive parts of the country you could ask for. I was even able to garner a ride for Helen in our BT-13 hack.

Oh, there were occasional dark moments. Accidents are bound to happen, some serious and some not. The fatal ones brought home reality with vivid clarity. My original Oakland check out pilot was right about one thing. The P-39 was very unforgiving of careless handling. Usually it was an easy airplane to fly, even the planned stalls were quite normal; but an accelerated stall, or unusual attitude stall could result in some wild unplanned maneuvers. One of these unplanned maneuvers was "the tumble." Sure, according to the Bell people's definition of a tumble it can't happen, but their definition of a tumble is far more restrictive than mine. To them a tumble means having the nose and tail of the plane rotate around the wings as it falls. Their test pilots had been asked to try to tumble the aircraft and, not surprisingly, they found that it couldn't be done. However, if an airplane falls like a rock and flops around in all manner of attitudes in the process, then to me that's tumbling. There is an aerobatic maneuver invented and named by a Czech pilot. Translated into English the name means "headache." and it closely approximates what the P-39 does in a tumble. The main difference is that most aerobatic airplanes tumble only briefly doing the "headache" while the P-39 seems to make a career out of it. It was not uncommon for a P-39 to recover from a tumble (my definition) by itself and fall into an inverted spin. It would then be possible for the pilot to very carefully pull it into an upright spin and make a normal, but gentle recovery. That is, if enough altitude remains. This whole process might take the better part of 8,000'. "Gentle" was the key word on the recovery. One of our less fortunate pilots went from 14,000' all the way into the bay — with fatal results, of course, — by trying to rush the spin recovery and going into one secondary spin after another.

I'm very well versed in aerobatics, both as a pilot and an observer. I know an inverted or upright spin when I see one, regardless of whether it's a straight forward spin or oscillating. My opinions about the tumbling aspect of the P-39 will not put the controversy to rest, I'm sure, but I've tumbled it twice and have watched others in tumbles at least five times, and as far as I'm concerned that's that. My own excursion into the world of tumbling came as a result of my weird sense of humor. It was decided by somebody that a good training exercise would be to have all three squadrons meet at the north end of the bay and engage in one huge dog fight. I say huge with reservations as none of the squadrons had a full compliment of airplanes and because of their age, maintenance was a problem. As a result, at least in the case of the 362nd, a squadron was lucky to put up eight airplanes at a time.

At any rate, we had this big melee going and I was hot on the tail of an opponent when he pulled up into a loop. At the top of the loop, I happened to look to my left just in time to see some poor soul fall into a tumble. I was so distracted laughing that I let my own airspeed fall too low and the P-39 promptly reminded me of my foolishness by snapping into an immediate tumble of my own. The stick jerked out of my hand and all I could do was keep my knees and hands out of the way of the flailing stick until the airplane decided to do something that I could recognize. Finally the ship stopped tumbling and settled down absolutely straight and level — but with no airspeed showing on the dial. I very carefully took a hold of the now quiet stick by two fingers and gently pushed forward to get the nose down. The fact that it responded showed that there had to be some forward motion. I hardly dared breathe as the nose dropped down enough to where the airspeed needle began to creep up the dial. I didn't do a thing more until I could see 200 MPH on the dial. only then did I re-enter the fray.

On another occasion one of our older more experienced flight leaders took six or seven of us newer guys up on a rat race. I believe it was Art Lingo. As usual I was tail end charley. He pulled out of a dive and went straight up into vertical rolls. With his experience, he knew the limits of the P-39 and rolled out at the top with minimum air speed. Several of the less experienced pilots didn't have his expertise and fell into tumbles just as I was pulling up. For a moment it looked like it was raining P-39s. I quickly scurried aside towards safer territory.

We couldn't have asked for a better C.O. than Major Egnes. he'd seen combat in the Pacific and knew what combat pilots had to go through. He even had two Jap planes to his credit. It was interesting to hear him tell about them. One was an unarmed reconnaissance airplane, but he said he was still shaking so bad from buck fever that he could hardly aim. Anyway, as a result of his experience, he gave us a long leash for our flying. He knew you couldn't become a polished combat pilot abiding by a bunch of safety rules that the enemy never heard of. Still, at this time the war seemed a long way away.

It was close enough though that somebody decided that we should go out to the rifle range and run through the qualifying course for shooting the G.I. .45 automatic. Carrying one on combat missions was going to be an option. I was a reasonably good shot with a pistol but remembering my experience in Alabama with the rifle, I kept my expectations low about qualifying for a medal, especially since I had heard that the .45 was a difficult piece to shoot. As in Alabama we weren't given any opportunity to practice, we were just handed the weapon and pointed in the direction of the target. At least this time we had good California weather and I wound up one bullet short of winning an expert medal. I like to believe that one of the bullets went exactly through a hole from one of the previous shots. It was hard to believe that one of the rounds missed the target completely in view of the tight grouping of the rest.

We were all intrigued with the Thompson sub-machine guns that were there, so we were given the opportunity to try for a medal with those too. Whenever possible it always pays to hang back and see what mistakes others make at something new. I could see that the biggest mistake the other guys were making was holding the trigger down too long and having the barrel climb skyward. By keeping my bursts of fire short, I was able to easily qualify for an expert medal with the Thompson.

Just about the only thing at Hayward that could have been classed as drudgery was the requirement to fly an hour each month under the hood in the squadron BT-13. This was supposed to maintain our instrument flying proficiency. A good trick, since many of us didn't have much of that to begin with. Two pilots would take turns riding shot gun for each other in the front seat for this less than exciting task. But we managed to have fun with it anyhow, at least I and the guys I flew with did.

There was a pear orchard on part of the field near the hard stand where the BT was parked. We would fill our pockets with pears and spend the hour "bombing" the lakes in the hills behind Hayward from about 3,000′. It was frivolity I would learn to regret in the months ahead. But then, since we were not allowed to fly if the weather looked the least bit threatening, it was hard to imagine that we would ever need the skill. The skills we developed were the ones we practiced and enjoyed the most. We spent many hours dive and skip bombing, and on aerial and ground gunnery.

Aerial gunnery definitely had it's down side. We used live ammunition and fired at a towed target. We took turns towing and, of course, had plenty of time while up there to wonder if some budding ace was going to get carried away.

I had some very scary moments while towing the target one day but it was not from that cause. I had taken off and was heading towards our target practice area which was about five miles out over the ocean west of the San Francisco Peninsula. As I crossed the peninsula, the target and cable just dropped off for no apparent reason. I circled the falling target and was horrified to see it heading right for a school yard full of children. It landed right on the front steps. I was heart sick to think that it might have killed who knows how many kids sitting on the steps. I rushed back to base and reported the possible catastrophe. A quick call was made to the proper authorities and the answer brought a flood of relief. "Oh, no problem, a teacher has taken it inside the classroom for show and tell." Whew!

I loved firing the P-39's 37mm cannon. It was supposed to be fully automatic like a machine gun but invariably jammed after firing two or three rounds. In spite of that I once destroyed a ground target with just two rounds. The target was about 25 feet square and built to last but apparently both rounds hit vital structures and it collapsed like a house of cards. That canon was murder on ground targets — a great morale booster! Trying to hit the aerial target was a different matter entirely. Our ammo was tipped with paint with a different color for each pilot so hits could be credited to the proper pilot. To the best of my knowledge no one was ever able to hit the aerial target with the cannon. Once, I was absolutely certain I'd hit it. Each round had a tracer element that made it easy to follow its path. Again I was only able to get two shots out of the cannon but was sure I saw one of them hit the target dead center and reappear out the other side. My hopes of being the first pilot to hit the target with the cannon were dashed back at base, however, when no trace of a cannon hit could be found.

The fact that the cannon jammed so easily made trying for a hit on the aerial target much more difficult. It was frustrating since I loved shooting it so much. When it jammed one day, I was so determined to get it un-jammed that I gave the charging cable a mighty pull and wound up with about two or three feet of cable in the cockpit. It's lucky no one ever had to depend on the cannon alone for doing battle.

P-39. Bell Publicity Photo.

We engaged in a lot of mock combat. Diving in a mock attack on an unsuspecting victim below you was called a bounce. I believe the expression came from the English. Apparently it came from the fact that after many of their diving attacks they climbed right back up, sort of like a bouncing ball. While we flew around anything else in the air was fair game. It didn't matter what or who it was — they got bounced! If they took up the challenge, the fight was on. If not, you just chalked it up as a clean "kill".

Hayward was just south of the Alameda Naval Air Station. This base was home to a lot of Navy pilots training in Douglas SBD dive bombers, the same SBDs that were being built back at Douglas when I worked there. The Navy had their dive bombing target in the south end of the bay. The target was a large wooden pyramid about the size of a modest house. The bombs the Navy used were little six pound practice bombs and when they hit the target (which wasn't all that often while I watched them) they didn't do much damage. Since their target was right in the middle of our stomping grounds, we frequently found them in the air getting ready to dive on the target. They were fair game and we never failed to make our presence known. They were a stuffy bunch and did their best to ignore us. I never in the world ever expected to see them under such circumstances!

We did a lot of bombing practice too, both dive and skip bombing. The P-39 had a fairly high wing loading and consequently was a very stable gun and bombing platform. We didn't have a sophisticated dive bombing sight. We just dove as straight down as was practical and when the target was steady on the nose we let the bomb go and it invariably hit with good accuracy.

We used a glass ball full of aluminum powder for our target. We took it out to the same area that we used for aerial gunnery. With our car type roll down windows, it was a simple matter to chuck the glass ball overboard. When it hit the water, the glass broke and the powder spread out in a nice visible circle. If it hit some part of the plane and broke, we were then without a target. We dive bombed it with our single 100 pound practice bombs and then proceeded to shoot at it with our guns until the powder disappeared. One day there were several fishing boats near our chosen spot, but since they were a safe distance away, we chucked the glass ball out anyway. They must have had sharp eyes and saw the ball hit because they all turned and started heading directly for the target. We quickly made our dive bombing runs and then fired away with our guns. You never saw fishing boats reverse their courses so quickly!

The Navy soon got grumpy about our bouncing their formations of SBDs and even called down to complain. Pretty stupid I thought. They were bound to be bounced in real combat, so they should have been happy with a little pseudo realism. It seemed to some of us that if they were going to be that small minded, we were more than justified in using their target when our glass ball didn't make it to the water.

So the next time we were without a target we used theirs. It was a little surprising, and satisfying, to see pieces of wood fly when we hit it with our hundred pounders. We continued to hit it until there was nothing left but floating bits of wood. The Navy obligingly rebuilt it, probably pretty self satisfied that their SBD jockeys were getting so accurate. It wasn't long until we destroyed that target also. Being completely devoid of modesty, I must say we were very accurate with our bombing. Getting three direct hits out of a flight of four was not uncommon. Even so, when the Navy's target was broken up into smaller pieces, it was easier to resort to skip bombing to finish it off. The Navy finally gave up on the pyramid concept and put up a ring of white rope supported with small buoys around the perimeter. It was far more difficult to hit the rope and even when we did, it merely bobbled around and re-surfaced. The Navy finally won a round.

Our friction with the Navy finally came to a head when our flight bounced them just as they were getting ready to peel off for their bombing run. We were loaded with bombs ourselves so we rolled over after the bounce and, probably unwisely, proceeded to let fly at their target. Since the bounce had already irritated them, they no doubt were galled (and enlightened) beyond description to note that our flight of four out-scored their whole squadron. In fairness to the Navy, I must admit that we did carry our dives to a much lower altitude than they did. This made accuracy somewhat easier to achieve. The Navy must have really burned up the phone lines because Major Egnes finally laid the law down. Leave the Navy alone! He was very emphatic!

Egnes did see the value of the Navy's six pack of small bombs. They gave much more practice per flight than our one, so he had our planes converted. The change made it practical to build a skip bombing target right on the field. It was probably the size of the wall of a house.

If any airplane was built for buzzing, the P-39 was it. It was stable, fast, maneuverable and the visibility over the nose was hard to beat. These were all features that made skip bombing a cinch. Some of us were confident enough in the skill of the squadron to walk out and stand right by the target to see how close to the center of the target the other guys were hitting. The down side to this was the inevitable buzz job while walking to and from the target. We all loved to buzz, and the lower the better. When you saw one of our guys coming, you had better dive for any depression you could find! More than once, I could feel my shirt lift off my back as some joker flashed inches over me at about three hundred miles an hour.

While motoring around at about 4,000′ looking for mischief one day, I looked down to see one of our guys buzzing the water and leaving a very respectable wake. We all were very competitive and I figured if he could do it, I could do it. Leaving a wake on the water was quite a challenge. The water had to be very smooth and you had to fly very low. As any seaplane pilot will tell you, judging your height over glassy water is quite a challenge. I suppose being close to shore helped with the effort. Anyway, pulling up to see my own wake was satisfying.

When you're twenty two, image is very important. So it was that I got the wild idea that I was going to look a whole lot sharper in cowboy boots. Looking back I'm amazed that Major Egnes put up with me. Cowboy boots and an officer's uniform really didn't go all that well together. Thank goodness my flying was more important to me than my image. The first time I flew the P-39 with them on was the last time. No foot wear could have been designed to be more uncomfortable to fly in than cowboy boots. I'm sure Egnes heaved a big sigh of relief when I abandoned them.

In addition to training, the squadron was also part of the Coastal Defense System. Two P-39s had to be parked on the end of the inactive runway fully gassed, armed, and kept warmed up by the ground crew. They were ready for an immediate takeoff at the sound of the alert siren. Two pilots were required to stand by in the pilot's room. When the warning sounded to scramble, they were rushed out by jeep to the planes that had already been fired up by the crew chiefs. After taking off, the pair of fighters would then be directed to the unidentified aircraft by ground control. The job of being on call was rotated amongst us pilots.

We were scrambled fairly often but invariably the bogeys turned out to be an errant C-47 or such. No doubt the button pushers knew this ahead of time on most occasions, but it was good training and a lot of fun anyway. Being a competitive bunch, it was always fun to see just how fast you could get in the air. I found that I could hit the throttle as soon as I hit the seat, close the door, strap in, plug in, and be ready to rotate just as the P-39 was ready to lift off. Only in something as easy to fly as the P-39 could you get away with something like that.

On a couple of occasions, we were scrambled at night. The button pushers must have thought they had good reason, but it sure left me puzzled. Just what was a day fighter going to do on a pitch black night? And why did they have all of the search lights waving around the sky, especially when the light operators had orders to shut off the lights the instant the light hit an airplane? I suppose that last bit was to keep from light blinding us — and we appreciated that — but how could they know we were not the enemy planes they were looking for?

After stumbling around for a while, it would be obvious that it was a false alarm. Always willing to find a way to enjoy myself, I'd spend the rest of the time dashing around trying to put out the lights until recall. It was more difficult than it might seem since the lights were waving around so frantically.

I wish one of the old hands had given me a little briefing before my first night landing in the P-39. It could have saved me a few skipped heart beats.

I dove into the pattern with the usual verve that I had acquired and as I was turning from base onto final, I lowered the retractable landing light. All of a sudden, the ship started to shudder and a wing started down — classic symptoms of the beginning of a stall! It couldn't be! I had all of the numbers right! A stall from this altitude would have meant certain death. Fortunately, the symptoms stopped as quickly as they had started. I soon realized that as the landing light began to extend it was able to give a slight aileron effect to the airplane as well as a slight shudder from the turbulence it created. Both symptoms, though slight, were magnified by the tenseness of my first night landing — enough to give me quite a start!

I loved the landing pattern that we had, at least in day light. The breeze was invariably out of the west so we started our run east of Hayward over the hills and dove down towards the runway. We planned our angle of descent so that we were right on the deck over the near end of the runway. At that point we pulled up into a tight left hand 360 that was almost a leaning loop, dropped the gear and flaps and curved around to land on the runway.

I liked to watch the other guys when they came in, just to pass judgment on how hot a pilot they were that day. A keen feature of the P-39 was the great streamers that came off the wing tips in high angles of attack, such as in a tight turn or pullout. A hot pilot could pull streamers about half way around the turn of his landing pattern. I lucked out one day and still had a trace of streamers almost three quarters of the way around. But then, perhaps it was just a high humidity day. Anyway, I had come a long way from my early sissy flights.

It was getting ever closer to the time we would be leaving good old Hayward and Egnes decided we should all have instrument tickets before we left. Our check rides would be done in the P-39. We had one with a hood that could be pulled down and it blanked out the whole world in the process. The idea was to climb to altitude on the wing of a flight leader who would be a combination safety pilot and an instrument check pilot.

Cal Williams was tagged to put me through the routine. We climbed out towards the Sacramento Valley where the sky wasn't so crowded. At about 5,000' we leveled out and Cal gave me the word to drop the hood.

It had been quite a while since I had made any effort at flying under the hood, so finding the world blanked out from me in the cockpit of this very responsive P-39 was intimidating to say the least. After a shaky bit of straight and level effort, Cal suggested doing a 180° left turn. I was about half way around the turn when I felt so disoriented that I popped the hood. "Well, that wasn't too good, get straight and level again and try one to the right." At least he wasn't yelling. Luckily I managed to get around the 180 to the right without spinning in. Cal was quick to find my skill adequate. "Great! Hey, there's the Marysville Buttes down there. Let's go down and buzz them. I've always wanted to see them up close." I had my G.I. Instrument Ticket. I may have been a lousy instrument pilot, but I sure knew how to buzz!

There was one other task to perform and that was a cross country. It was decided that it would be with a flight of four planes with each pilot taking a turn at leading the flight over a portion of the course.

The destination decided on was San Diego. Map reading was another area that I lacked experience in. I had trouble matching what I could see on the ground with someone's idea of what it ought to look like on a map. In due time, it all came into focus; but during this period, I was still very much the beginner. Naturally they had to pick me to lead off the first leg and I promptly headed off down the wrong valley. Luckily somebody in the flight knew what they were doing and herded me in the right direction.

It was a very interesting trip anyway. The range of the P-39 is such that we had to stop at Bakersfield for fuel. From there we were able to make it on to San Diego. When we arrived over the area where the airport was supposed to be located, I couldn't see it anywhere. There just wasn't any airport in sight. I can no longer remember the name of the flight leader who led the flight but he circled what appeared to be a mixed housing and industrial area a couple of times and then proceeded to peel off and land on one of the streets! I was awe struck, but as he rolled out with no difficulty, it finally dawned on me that Lindbergh Field had been thoroughly camouflaged and they had done a very good job indeed. Even as I touched down, I still had my fingers crossed and didn't feel completely at ease until I was parked next to the others. Obviously our flight leader had either been there before or had been thoroughly briefed. There was no way that I would have landed there had I been alone.

I was amused at one of the local G.I.s who obviously was fairly new to the service. We were due to head back without much delay so I just left the door wide open when I got out of the plane. I was leaning against the wing when this G. I. came over, saluted, and looked the plane over. He came over closer to me and asked: "How many people does it hold?" It seemed like a pretty dumb question since the door was wide open and it was plain to see that the thing was just barely big enough to hold a pilot. But, he seemed to be sincere, and I've always tried to be gracious and never threw my rank around, so I politely gave him a tour of the airplane. Eventually we headed back to Hayward. It certainly was interesting to cover all of that familiar country at a ground speed of better than 250 miles an hour.

I had my first close up look at a P-38 while I was motoring around the north end of the bay one day. He came by on a southerly heading and was in a perfect position to bounce. I resisted the temptation and just closed in on him to get a good look since I hadn't seen one before. The pilot didn't see me so I tightened up the formation. He was obviously on a cross country trip because he kept looking down at a map, then over the left side of the ship at the ground and back again at the map. Since he was not being very aware of his surroundings, I decided to tuck in very close and watch the surprise on his face when he finally looked my way. I over lapped our wings a good five feet and waited — and waited. We flew all the way down to San Jose and he still hadn't looked my way. Why he hadn't at least seen me out of the corner of his eye, I'll never know. I finally grew tired of waiting and was tempted to speed up and cross in front of his nose to wake him up. But, the insignia of a major was plain on his collar so I just prudently peeled away and left him.

A couple of days later I was once again tail end charley on a local flight. Low and behold the SBDs were out in force — and were in a perfect position to bounce! My conscience told me to hold back but you've got to stay with the flight, right?

There were big white buzz numbers on the side of the fuselage and under the wings of the P-39 that were supposed to discourage just such frivolity. Being a cautious soul, I did a very tight barrel roll around my chosen "victims", giving them the only view of the P-39 that didn't show a number, the top side. Two of the other guys were not that cautious. They were nailed. Evidently the stuffy Navy Commanders had had enough and called down through channels demanding court martials. We no sooner landed than the flight was ordered to report to Egnes. I walked into the office right in the middle of the chewing out the rest of the flight was already undergoing. It was a beauty. Egnes was livid and turned to me. "Mace! I know you were on that flight. They didn't get your number but I ought to let them have you anyway! Get out of here!" I got. I'm not sure how serious they were, but the word was that court martials were in the works for the two who had their numbers taken. In any case, it all seemed to be forgotten after the move from Hayward.

There was a feeling of sadness on my part when we left Hayward. Those were awfully good times we had there. But the feeling was countered somewhat by the adventure that a move to a new place inevitably brings. Helen had to be left behind long enough to sell our Harley Davidson motorcycle. She knew how to drive the big Harley but at less than a hundred pounds found it impossible to kick the thing to life. The floor of the warehouse where our apartment was located was at truck-bed height and had a ramp leading from the floor down to street level. Without me the only way she could start the bike was by coasting down the ramp and letting the clutch out when she hit the street. She handled the situation fine and was able to sell the bike.

We hated to part with it, there were a lot of good memories associated with the old Harley. Since gasoline was rationed, I used to occasionally ride it out to the hard stand where the P-39s were parked and casually mention to the crew chief that it sure would be nice if the tank was full when I got back from flying. Sure enough it would be. One hundred octane was not the easiest thing on the valves, but it was better than nothing.

Before the Harley we had a light weight Triumph which was a bit on the small side. on a week's pass, we drove the old '33 Ford down to Los Angeles and while we were there we found the good buy on the Harley. Getting both vehicles back to Hayward might have been a problem except that Don Rice also lived in the Los Angeles area and had ridden down with us. We managed to talk him into driving the Ford back while Helen and I had a real nice trip back through Yosemite on the Harley. Don was not so lucky, he had a flat on the way back and was disgusted to discover that we didn't have a jack. To this day he won't let us forget his miserable trip back when we get together at our reunions!

After selling the Harley, Helen drove on to Pocatello in the good old '33 Ford two-door. We eventually traded it in on a cherry '39 Ford coupe in Pocatello and the old '33 ended up in the happy hands of a local Indian. Once more the three squadrons of the 357th would be scattered over the country. The 363rd went to Ainsworth, Nebraska, the 364th to Casper, Wyoming, and the 362nd found itself nestled amongst a gang of bombers at Pocatello, Idaho. We all had the feeling that our stay in Pocatello would not be a long one, and that combat would not be too far in the future. Time would prove us right.

In the meantime, Helen found yet another place to stay. it was a motel with a kitchenette and served us very well. At least two of the other married guys also located there, including Dave Perron and his wife Harriet. At least Helen would have some company while I was busy at the base. Although Pocatello wasn't exactly Hayward we did find plenty of things to do in the area like duck hunting and horse back riding. We also decided that Helen should begin to learn to fly herself. Since Pocatello was well away from the coast, private flying was still allowed there. There was an active little airport on the way out to the base so Helen started taking flying lessons in a three-place Stinson 105. We didn't stay in the area long enough for her to solo, but it was a start, anyway.

Our time in the P-39 was spent in more gunnery practice as well as mock combat or whatever else we could come up with. It was on a ground gunnery mission that I almost joined the Caterpillar Club. The crew chief of the P-39 that I was going to fly asked me to empty the drop tank while I had it up as he wanted to remove it when I returned.

There is no fuel gauge for the drop tank, all you can do is go by time in the air. I asked him if the tank was full and he assured me it was. A full tank would have meant a minimum of an hour of flying on the tank.

Right after takeoff, I switched the fuel valve over to the drop tank. Fifteen minutes later I was right on the deck bearing down on the target at 275 MPH when I was suddenly greeted with total silence from the engine. I immediately switched the fuel valve over to a main tank and turned on the booster pump. Instead of the smooth purr of the engine returning to life, there was nothing but more dead silence. The gunnery range was located out in the boonies on some extremely inhospitable lava beds. The lava formed house-sized lumps and crevasses. Trying a belly landing on that kind of surface would have meant sure disaster. I decided the only thing left to do was to zoom up until my airspeed was down to near stalling speed and bail out. I salvoed the drop tank and pulled up. As soon as I was nearly out of usable airspeed, I unbuckled the safety belts and started to open the door to bail out. Almost as an after thought I decided to check the altimeter — I had no more than 400'! We had been warned more than once not to try bailing out at less than 500'.

I didn't have a whole lot of time to mull over my options. Neither bailing out at less than 400' or belly landing on that gruesome surface held much appeal. Since two of our guys had survived horrible crashes earlier, I decided to buckle back in and take my chances on a belly landing. Just as I was about to touch down, the engine caught and started running on about six cylinders. It wasn't much, but it was enough to keep me out of the lava beds long enough to stagger over to a nice stubble field where a safe belly landing could be made. I dropped the flaps and was just about to cut the switch when the engine started running on all twelve cylinders. What a beautiful sound that was! When I returned to base, Egnes not only showed little empathy for my narrow escape but moaned about the loss of the stupid drop tank. I wonder if he ever thought of that overseas where we dropped up to 100 tanks a mission, often with considerable fuel in them.

There were several bomber types on the field, but B-24s outnumbered the rest. One day near the start of the day's activity, an orderly came over to the pilot's room with a note from one of the bomber pilots. In effect it read:

"If any of you fighter jocks think you can beat my B-24 in a dog fight, meet me at 5,000' over the gunnery range in 20 minutes." It was a challenge that would not be missed. We all made a dash for one of the few P-39s available. My sprinting ability still stood me in good stead, and I managed to make it to one of the fighters. No effort was made to form up into any semblance of an organized formation, we just wanted to have at that B-24. Eager Beaver Alvin Pyeatt was just ahead of me as we dove in on the attack. He was so eager that he pulled on the stick a little too enthusiastically during the pullout and did two or three snap rolls so fast that I couldn't count them. Luckily he had enough speed to recover rather than falling into a tumble.

The guy in the B-24 didn't stand a chance of out maneuvering our P-39s, of course, but he put up such a whale of an effort that we were all pretty awe struck over his ability. After the flight was over we invited him to a get-together. He turned out to be a huge bruiser that any NFL team would have been happy to have as a line backer — and a heck of a nice guy to boot. He invited us all up for a ride in the B-24 where we each took a whack at flying his monster. It was only then that we could really appreciate the job he had done with that B-24. I think somebody managed to sneak him into one of the P-39s (I'm sure the only way he would have fit was without a 'chute) and turned him loose.

Our P-39s were kept on a huge ramp area adjacent to the taxi way that ran from where the bombers were located to the end of the runway.

On one morning that I was scheduled to go out on a flight as tail end charley (as usual) , the airplane that I was to fly was parked the farthest away from the taxi way and naturally I was also the last one to be dropped off. This meant that I had to hurry to join up with the rest of the flight at the runup area. Just as I started moving out, a B-25 appeared on the taxi way and he would have beat me to the narrow part of the taxi way had I just proceeded at a normal pace. If he had beat me, I would have been in for a long wait while the pilot went through his twin engine runup routine and my flight would have been long gone — so I picked up the pace.

And he picked up the pace. We both were determined not to be beaten and after several of these countering moves I had enough speed that I think I could have made a takeoff! Anyway, I won. I didn't think a thing more about it until after the flight and I walked into the pilot's room and there was Major Egnes hollering at Pyeatt for that wild exhibition of speed on the ramp. He was really livid! Evidently there must have been someone of fair importance aboard the B-25 that had given Egnes a chewing out. It sure made the usually docile Egnes get into a real stew.

We all were confident, cocky, hams that loved to be flamboyant, but Alvin Pyeatt had the edge on us. Enough so that the recruiter people picked him to be on a poster looking off into the "Wild Blue Yonder." He and Egnes had also become pretty buddy buddy, which made this scene all the more surprising. My first impulse was to just disappear, but my conscience wouldn't put up with that. I had to wait for Egnes to pause for a deep breath before I could even butt in. "Major Egnes, I don't think Pyeatt is the one you want, that was me speeding on the ramp." "I know, I know, I saw you going too fast too, but it was Pyeatt that was really showing poor judgment!" and he went on with the verbal castration. There was nothing more for me to do but quietly slink away.

The time at Pocatello was coming to an end. Although strong rumors were that England was our destination, there was not a hint of what we might be flying. We were positive that it would not be the P-39, however. We would leave it behind along with the memories. As our group historian, Merle C. Olmsted, reports in his book "The Yoxford Boys", there was a discouraging number of accidents during our period of training. Our three squadrons would leave with a record of frequent accidents and the saddest of all our stats was that we'd lost thirteen pilots. Most of them were for mechanical reasons like engine failures and jammed landing gears. F/O Yeager would account for two of these in one five day period. One was a belly landing caused by a stuck landing gear, and the other was a bail out from a burning P-39. They were the first of many emergencies Chuck Yeager would have to face in the future.

Unfortunately, too many of the accidents were caused by pilot error, which was understandable since a gang of pilots with not much more than 200 hours in the air were rushed into a complex fighter with an unforgiving nature. Not all of the "pilot error" accidents could be blamed entirely on the pilots however. When the early models of the P-39 were delivered, the flaps and the landing gear were both activated by identical G.I. switches placed side by side on the instrument panel. This arrangement just begged for a pilot error accident to happen. As a fix, a little plastic shield was placed over the landing gear switch which helped, but it was not the sure cure that was needed. Later models had one of the switches placed under the seat. At least we were leaving for combat as a bunch of seasoned fighter pilots. As junior member of my squadron, I wound up with close to 250 hours in the P-39 while our more experienced leaders had as much as 700! This gave our squadron a very healthy average of close to 400 hours per man.

This next move meant leaving Helen behind. She went back to Los Angeles to live with her mother and did her part working for Lockheed Aircraft in the Stenopool. Leaving her behind clearly brought home the reason the Air Force didn't care for their fighter pilots being married. It certainly made going off to combat tough indeed. I would spend many a lonely night during the next year and a half. My heart was definitely back in Los Angeles.

Early in November, 1943, the entire group found its way to Camp Shanks, New Jersey. This was our jumping off point for boarding the Queen Elizabeth, which carried us safely to Glasgow, Scotland. A new and significant chapter in the life of the 357th Fighter Group was about to begin.

COMBAT AND THE P-51

We left Glasgow by train for Raydon Wood Airdrome in the county of Suffolk, England. Getting used to the weather, which contrasted heavily with what we had in Hayward, would take some effort. It was drizzly most of the time which made the half finished base a sea of mud. Our group had been assigned to the 9th Air Force whose duties were primarily ground support. In view of the type of training we had, this seemed reasonable. The arrival of the group's first airplane did little to satisfy our curiosity about what we would be flying, though. It was a Piper L-4.

A few days later our curiosity was satisfied to everyone's great satisfaction with the arrival of our first fighter plane. It was a North American P-51-B Mustang! I was about to lightly cross the path of one of my prewar racing heroes. The well known Art Chester had been put in charge of developing the engine, cowl, and spinner installation on the Mustang. Never in my fondest dreams did I ever imagine that I would have the chance to fly such an airplane! The big event of the first flight, when it happened, must have been overshadowed to a large extent by all of the hours flying the P-39 because I actually have no memory of that first flight. Almost every detail of my first flight in the P-39 is vividly etched in my memory, but not so for the P-51. This seems very strange in view of the great differences between the two. However, Bud Anderson, one of our group's highest scoring aces, reports the same phenomenon. I do have memories from my early flights, however. One of the strongest comes from my first buzz job — which could have been on my first flight. When I "hit the deck" , I was surprised at the bouncing around the '51 was doing from what seemed to be a modest amount of turbulence. The P-39 had always been so rock solid.

A later photo of our beloved P-51s, carrying drop tanks and painted with black and white "invasion stripes" heading for another mission over Europe.

After getting a couple hours on the P-51, I was in a pretty good position to compare the two airplanes. They both were easy to fly, and though the tricycle gear on the P-39 made takeoffs and landings a breeze, the tail dragging '51 was not all that hard to handle either. There were two major advantages the '51 had over the '39: range and altitude capability. The P-51 could comfortably operate at twice the altitude of the P-39. The two-stage blower on the Mustang accounted for most of that difference. The biggest drawback in the '51 was how blind the pilot was while taxiing. To prevent ground collisions, it was common for some base Commanders to require the crew chiefs to ride on the wing while taxiing. Though the cruising speed of the two airplanes was not all that different, the '51 was cleaner and capable of picking up much more speed in a dive. They both were liquid cooled and this made the airplanes very vulnerable to ground fire of almost any caliber. This drawback, and the phenomenally great features the '51 had for high altitude long range escort was finally brought home to the powers that be.

Late in January, 1944, the 357th found itself trading bases and missions with a P-47 group based at Leiston. The P-47's air cooled engine and tank-like structure made it admirably suited for ground attack work. It also meant trading commands, we were now the first Mustang group to fly for the 8th Air Force. Our main mission would be to escort heavy bombers on long range missions where allied fighter planes had never been able to go before. The Mustang was capable of accompanying the bombers to any target the bombers cared to hit. The bomber crews would soon be afforded a good measure of protection they had previously been denied. Time would show that the move to escort duty had not only been a blessing to the bombers, but to our group as well. Though the 357th would suffer heavy casualties in the months ahead, the number would be considerably less than had we been exposed to constant ground attack missions in the vulnerable P-51.

Our group had the distinction of having two brothers numbered among its members. This may have come about because there was a third brother in a fairly high position in the Pentagon. Captain Lloyd Hubbard, the older brother, was a flight leader in the 363rd Squadron and Lt. Glenn "Hub" Hubbard was a tail end charley member of the 364th and the bane of the Squadron leader. Some of his exploits that earned him this distinction will be related later.

Capt. Lloyd Hubbard was generally regarded as the hottest pilot in the group. Certainly he was as good as they came. He and several others were picked to fly a mission with another group, the 354th, to become familiar with combat flying before leading us on our first mission. He flew this ground attack mission on the 9th of February and was killed by ground fire. A more disturbing way to have the hazards of combat and ground attack missions brought home could not be found.

Our new base was always referred to as Leiston, the name of a nearby town. Yoxford was the name of another nearby town. Our deployment to England had been done under the usual umbrella of secrecy, so it was a bit of a shock to tune in to the German propaganda station on the first night at Leiston and hear "Lord Haw Haw" welcoming the "Yoxford Boys" to the area — some of us by name! After a number of weeks of combat, it was satisfying to hear his jovial tone of welcome change to a more bitter reference to us as those "Dirty Yoxford Boys"! It would be a few weeks at Leiston before we could begin flying combat, though. We had yet to receive our full complement of airplanes. They were being off loaded and assembled at Blackpool and would have to be ferried to Leiston.

Since practice flying in the '51 was a little hard to come by, I volunteered to do some of the ferry work along with Tom Beemer. Being very good looking and single, Tom had wasted no time in acquiring a very pretty little Scottish girl friend. I had the opportunity to meet her at Blackpool. She sure was a cute little charmer. Seeing the two together just made me more homesick for Helen, however. Earlier I'd had to "celebrate" our first wedding anniversary on the Queen Elizabeth without her.

Ferrying the airplanes to Leiston meant flying across England at just about its broadest point, which took all of about an hour. Getting to Blackpool was another matter entirely. You couldn't just take a train from one city to another in England. Almost all railroad tracks lead to London. What should be a 100 mile trip could end up being 200. At least the train ride gave us a chance to see the country side.

In order to make the flying trips a little more interesting, I decided to make them purely on dead reckoning. In other words, by not referring to map or check points and relying on compass and time alone. The time factor was academic since Leiston was on the coast. When I came to the water, I'd know I had gone far enough. To avoid any temptation to look for check points, I buzzed at tree top level all the way. It not only was fun but gave the added advantage of keeping me below the usual bad weather. I had never been a fan of dead reckoning. It's impossible to counter all of the possible variables. In this case, however, the shortness of the trip precluded getting too far off. I even hit Leiston right on the money a couple of times. The P-51 was really a great flying airplane, it would sure be hard to beat. I only wish that I had been given the opportunity for a lot more combat type of practice flying instead of just ferrying the group's planes into Leiston.

There was one hazard that didn't show up in any of our instruction books, the damned British barrage balloons. Most of the larger towns had them. They were anchored in herds on cables and were supposed to be kept just below the cloud level. Given the poor visibility that invariably accompanied the usual low ceilings, flying below the clouds made them quite a hazard. In an effort to ease the threat to allied pilots, each herd had a low power transmitter sending out a steady tone that got louder as you neared their area. Unfortunately the tone volume increased the same whether you were headed right at them or were just missing to one side or the other. Consequently when you picked up the increasing volume of the tone, it left you with the dilemma of choosing to turn right or left or just forging straight ahead in hopes of avoiding them. At least the range of the transmitter was short enough that you didn't have long to worry. I finally decided that just plowing straight ahead with a sharp eye out was as good a choice as any. Fortunately I was always right with my choice and it was extremely comforting when the tone volume began to decrease. Obviously flying above the clouds would have eliminated the problem, but that is not always practical and you do have to come down sometime. I can't vouch for it personally, but it has been said that more barrage balloons were lost to American gunfire than anything else. One joke that always made the rounds was the threat to cut all of the balloons loose and let the island sink! It helped to counter the often heard British complaint that we Yanks were over paid, over sexed, and over here!

Eventually enough airplanes were delivered that each of us was assigned a specific plane. We wasted no time in naming our charges. A few pilots, or possibly ground crew, even added paintings on the noses as well. I think the hands down winner in the nose art division was Arvil "Robie" Roberson's "Passion Wagon" which had an almost life size painting of a very seductive nude. The fame and publicity achieved by this airplane probably exceeded that of Robie himself. I chose a more conservative route and named my P-51-B after Helen. I painted the name "Sweet Helen" on the nose myself.

It would be February, 1944, before the 357th would fly its first combat mission. On the morning of the 12th, the Charge of Quarters rousted us out of bed. After struggling into cold clothes, some of us stepped outside to be greeted by the sight of a cold wet overcast that clung to the tree tops. We promptly jumped back in bed, nobody ever flew in weather like that. All too soon the C.Q. was back in the hut, "Come on you guys, they're waiting for you over at briefing, you better get going!" We couldn't imagine why we should, but then, maybe the brass just wanted to put us through the routine as a rehearsal for the real thing. Obligingly we sat through the briefing and found that the mission would have been an easy ice breaker sweep over the Rouen area of France. Easy as combat missions go, that is. We were soon pointed towards the 'chute loft and driven out to our planes. Man, this was carrying this rehearsal thing to an extreme!

Capt. Dave Perron, my first flight leader and good friend, who was killed in action April 1, 1944

I had never seen my ship, Sweet Helen, look so cold and uninviting. Casting a leary eye at the gloomy overcast, I climbed in, buckled up and plugged in the necessities. Soon the cold early dawn air was ripped by the crackle of idling Merlin engines from all points of the field.

Dread was really beginning to take over, this was looking less and less like a rehearsal.

I was really regretting that frivolous attitude towards instrument flight training we had back in the States. I'm not sure if anyone in the squadron was really very good on the gauges.

A typical four plane formation shot taken at a later time when the group was flying early P-51Ds. Front to rear; Capt. Harvey F. Mace, Lts. Weaver, Martinek, and Goss.

I was in Dave Perron's flight and he liked having me as his wing man. This meant that while he flew on instruments through the misty muck, I would have to fly extremely close formation to keep him in sight. Dave was an excellent pilot, as I think we all were, but like most of us, he was not all that steady on the gauges. Add to this the fact that we were in fully loaded planes, to which we were unaccustomed, and it was going to be quite a challenge.

Before long, pairs of Mustangs were racing down the runway and disappearing into the mist. All too soon it was our turn. I overlapped our wings just prior to entering the clouds and clung to Dave like a wet dish rag on a sink. As we climbed up through the muck he wobbled around a bit but eventually we made it through. What relief it was to break into that clear sky. But now, besides facing my first combat mission, I had to worry about how we were ever going to get back down through that mess and find the runway. It was a situation I had never faced before. If this weren't enough to make you break out the Aspirin, add the fact there could be literally hundreds of other allied planes trying to occupy the same air space... a modern traffic controllers worst nightmare. Amazingly — this time — we all made it through.

In the clear air on top, the group gathered in combat formation and headed towards enemy territory. This was it. The fun and games we enjoyed so much back at Hayward were over. Now we were playing for keeps. The mission for the day was just a sweep, or patrol, as it were. There would be no bombers to escort, nor would the penetration of enemy territory be very deep. However, it would expose us to some anti-aircraft fire and possible enemy aircraft opposition. If E/A came up, we'd fight. If they did appear, it could be from any direction, at any time. We reached our planned altitude about the same time we reached enemy territory. Our heads really swiveled and our eyes strained.

All of a sudden ominous black puffs were bursting around us. Anti-aircraft fire! Each black puff with its yellowish center of fire was sending out potentially lethal shards of metal in all directions. A sudden new meaning was given to the word combat, those people down there were trying to kill us! Mustangs twisted and turned to get out of the line of fire! Fortunately the volume of fire was light and not too accurate so no one was hit — but it sure got our attention. No enemy aircraft were seen during the mission and that suited me just fine.

Sweating over the weather and dodging anti-aircraft fire was enough to face on my first combat mission. Eventually it was time to turn and head for home and face the problem of getting down through that layer of muck and try to find the runways at Leiston. I was beginning to wonder which would turn out to be our worst enemy, the weather or the Germans. For all too many, it would be the weather.

As we let down through the overcast, I heaved a big sigh of relief when I found that the warming morning sun had raised the ceiling to acceptable levels. Enjoy, for time would show that it would not always be that way. Faced with the spectre of much bad weather in the future, I wasted no time in refining my own meager ability at flying on instruments. It was readily apparent that I may need it at any time. Sadly, not too far in the future, Dave Perron, my flight leader and my good friend, would once again be climbing through an overcast with a new recruit on his wing. They would disappear, never to be seen again. I would miss the times he tried to boost my morale when I was feeling the lowest over being away from Helen.

It was being forcefully brought home that being a fighter pilot was not all fun. Imagine reporting for work with a bulky pack on your back and being strapped into a confined cubicle that leaves precious little room to move. The tightly packed life raft that you sit on for the next four to seven hours will make the desire to shift your butt around very strong indeed, with no snacks, coffee breaks, or trips to the rest room. But, you haven't reached the hard part yet. Your job requires you to launch your cubicle into hostile skies at breakneck speeds and be shot at. Since it is very unpleasant to be shot at — or run into some other poor soul in similar circumstances — you have to swivel your neck constantly. This will be made as difficult as possible because of the mask, hose and wires that you must wear on your head. Even blowing your nose will have to wait for the end of the day. Add the possibility of getting lost, having to deal with the threat of vertigo in horrible weather, worrying about possible engine trouble and it just might make you wonder why you ever chose to be a fighter pilot in the first place.

But, it wasn't all doom and gloom. Having a machine of such beauty and power respond to your every whim was a rush, as was seeing the earth and the beautiful scattered clouds from angles few were privileged to see was a rush. Testing your ability to hold your airplane within inches of another was a rush. Skimming home base after a very tough day with a four hundred mile an hour buzz job followed by a victory roll was a particular rush. No, I wouldn't have had it any other way.

After Dave's loss, John England became my flight leader. John was a heck of a fighter — but a bit rough as a pilot. He had absolutely no faith in the artificial horizon for flying on instruments. He chose to rely solely on the old needle, ball, airspeed routine. It was very reliable but not nearly as easy to be steady with. Throw in John's somewhat slam bang approach to life and it made flying on his wing through an overcast quite a challenge. I've seen John approach a chair where the squadron cat was sleeping, grab the cat by the scruff of the neck and throw it over his shoulder and sit down without even bothering to see where the cat landed.

John seemed to be one of the very few pilots that had no fear or feelings. His seemingly total disregard for danger was displayed on one of our missions deep in occupied France. It was one of the very few ground attack missions our group would be assigned to do. In place of the drop tanks, we were loaded with bombs and were out to find targets of opportunity. When the group reached the target area, we split up and John and I wound up alone stooging around looking for a target. Eventually we came across an enemy air base. There were no aircraft on it but we decided to hit it anyway since it was the most inviting target we had found. John peeled off and I followed right behind. I was torn between hitting a hangar or going for the runway intersection. I opted for the intersection and let the bombs go. Unfortunately I had to rush to stay with John and didn't have the chance to see where the bombs hit. The base was later confirmed as the home of the Abbeyville boys, a tough bunch of guys.

Captain Calvert L. Williams. Besides the incident related in the text, he was also responsible for our squadron's first confirmed victory, an ME-109, on Feb. 20, 1944.

On the way home, any sensibly cautious pilot would have climbed to a safer altitude where you would be above most of the danger, or fly right down on the deck where you might take most of the ground firing people by surprise. But not John, he chose to meander over enemy territory at 5,000 feet as if he was on a Sunday sight seeing trip. I was nervous all the way home from flying at such a dangerous altitude. How we made it home with hardly a shot being fired, I'll never know. John survived the war in spite of additional tours and the destruction of 19 enemy aircraft. Unfortunately he met his end in France in a flamed out F-86 in 1954. England Air Force Base, Louisiana, was named in his honor.

Weather conditions sure varied over Europe. On some days the overcast would be so dense that even with overlapped wings, it would be a strain to see the airplane you were flying wing on. On other days it would be like flying inside an opaque ball with no ground, sky or horizon visible anywhere, and yet you could see another airplane several hundred yards away. It was on one such a day that I was flying on Cal Williams' wing, the same Cal that had given me my instrument flight check in P-39s back at Hayward. We had reached the enemy coast at about 25,000′, still in this opaque ball when the powers that be scrubbed the mission.

The group started dropping tanks and turning for home. I immediately dumped mine as we started the turn but Cal's didn't come off. Having the drop tank release mechanism get balky at times was not unusual. Soon he started porpoising to shake them off which was an accepted procedure. My immediate thought was that he must be a heck of an instrument pilot to be trying to shake his tanks off during an instrument turn! The gyrations that he was going through soon became more than I cared to keep up with. Since we were headed for home anyway, I just took off on my own. The last I saw of Cal was seeing him spiraling down towards a thin spot in the muck below us. When we returned I complimented Cal on his ability to shake his tanks off in an instrument turn. "Shake the tanks off!? Hell, they weren't stuck! I had vertigo and was close to spinning in!" Well, at least he was honest.

We had come a long way from those blue skies of California. I think the worst day was having to take off in a snow storm. There was just enough visibility to see a small bit of the runway ahead. The P-51s ahead of me were disappearing in the snow while still three point. For all I knew I might be just about to add to a growing pile of aluminum on the end of the runway. All that lay ahead was sweating out the instrument flying, the flak, the possible mortal combat and four or five hours of wondering how we were ever going to find our way back down when — or if — we ever got back. Luckily by the end of the mission, the weather had improved for us one more time.

Getting home from the missions in bad weather was an adventure all by itself. The group was frequently separated by flak, weather, dog fights or all three. We had a general heading from the target area that was designed to get you back to somewhere near England. If the weather was still bad at about 30 miles out you could call for a fix and get a fairly accurate heading to your home base. It was not uncommon to have the visibility so bad that you couldn't see your field until you were right on top of it. For those occasions, a mortar-like device fired a big red ball every few seconds over the base to help you find the field. If you were lucky, you would find yourself close enough to see the field without being right over the mortar when it went off. If you did, you had it made — providing you didn't collide with some other totally fatigued, wrung out pilot trying to get in. You can imagine the confusion caused by poor visibility and close to half a hundred pilots coming in from all directions trying to get down as quickly as possible.

The pitfalls of flying combat missions seemed endless. For instance early in my tour,

Ray Smith, my crew chief, and I enjoy some unusually good weather on the wing of Sweet Helen II. The date was September 15, 1944.

I settled down in the cockpit of my plane one morning and, as usual, my crew chief Ray Smith helped me strap in. I started my familiar preflight cockpit check only to have my hand fall on a totally unfamiliar throttle control. "What in the world is this, Ray?" "Oh,that's part of your new gyroscopic gun sight." No one had ever said a word about any new gunsights much less given a breath of explanation on what they were supposed to do. Nevertheless there, where my trusty old 100 mil sight used to be, sat a totally strange looking contraption. It was bad enough that I never had the opportunity for any gunnery practice in the P-51, now I was about to be sent out on a combat mission with a totally strange, complicated gunsight that I had never even heard of. I would just have to pick the brains of my crew as much as possible in the few minutes I had before taxiing out.

"How in the world does it work?"

"Well, you see that dial on the face of the sight with the little airplane symbols on it?" Yes, there on a dial were little figures of enemy airplanes and a pointer. "Well, when you identify the enemy plane that you're going to shoot at, you put the pointer on the appropriate figure. Now, if you'll turn the sight on, I can show you the rest." I did, and there on the reflector glass appeared a circle of six little diamonds in place of the 100 mil ring I was so used to — along with a pip that wasn't even in the center of it! "Twisting that new grip on the throttle will widen or narrow the diameter of the circle of diamonds. Now you adjust the circle to just touch the wing tips of the E/A. That gives the sight the proper range. Then all you have to do is put the floating pip on the target, pull the trigger, and you've got him." Oh really! "Suppose all of this fancy stuff gives up the ghost, then what do I do?" Well, just flip this switch here and the fancy stuff is replaced with a fixed ring and pip like the old sight had."

I flipped the switch and sure enough a fixed ring and pip appeared. The only trouble was that it was a miserably small looking ring compared to the 100 mil ring that I had practiced so much with. Well I was stuck with the sight so I ran through the drill as I taxied out. First, I would have to identify the E/A and hope that he didn't do anything rash until I put the pointer on the proper symbol. Second, the enemy pilot would have to present me with a view of his wing span and hold it steady at my speed long enough to close the diamonds down to his wing tips. Then if he held his maneuvering down enough to put the floating pip on his airplane, I would have him. Hmmm. We took off and on the climb out, a couple of Mustangs crossed my nose which gave me the opportunity to try putting the wandering pip on them. Even under these stressless conditions, putting the pip on a moving target was far from simple. I just chose to switch on the back up ring and pip and wished I had my old sight back.

You had to harden yourself to the tragedies of war if you were to survive with any degree of mental stability. You just couldn't tear yourself emotionally apart over every tragic, traumatic thing that happened. I've watched many B-17s blow up with no survivors, and many that fell mortally wounded with only half of the crew escaping. I've seen bombs wipe huge corridors of destruction right down the middle of cities. I've flown along side troop carrying C-47s that went in with all hands. I was flying off the left wing of one when the left engine area caught fire from a flak hit. The fire grew quickly as the frightened troopers began bailing out. The first four got out in rapid order as the fire spread to the cabin area. The fifth man came out trailing smoke. The sixth man came out completely enveloped in flames which had also eaten through his static line. There would be no 'chute. It was probably just as well since he couldn't have survived the burns for long. All of the rest died with the airplane.

Since our field was right on the coast, it was a haven for many distressed combat planes looking for the nearest friendly place to put down. On two occasions, I've stood within a couple hundred feet of belly landing P-47s. One plane was heavily hit and the pilot so badly wounded we had to lift him out of the cockpit. Several times I've watched crippled B-17s coming in on their final approach shooting red flares. This was the signal to have ambulances ready as there were wounded on board. The list goes on.

The space occupied by a typical force of American bombers and fighters on a mission would be hard for a layman to imagine. The stream of airplanes could stretch from 50 miles for a small force to 150 miles for a really large mission. By necessity we fighter pilots had to give the bombers a wide berth. If you got too close, nervous gunners would shoot and they had every right to. As a result, fighters on the right side of the bomber stream might engage E/A in a dogfight while the fighters on the left would never see it. This was especially true if visibility was poor, as it often was. The bomber stream was composed of "boxes" of bombers. The boxes were formed by squadrons and groups of bombers that bore little resemblance to a box. I don't know the origin of the term, but it served the purpose.

There were variables of course, but generally a fighter squadron or group would be assigned to cover one box of bombers. This neat order of things could be easily scrambled by the introduction of flak, enemy fighters, or weather. Because of it's increased range over other fighters, the introduction of the P-51 drastically reduced the number of bomber losses to enemy fighters, but we couldn't be everywhere all of the time.

The agony that bomber crews had to endure was almost unmatched during the war. They deserve more than they could ever receive for their sacrifices. I'm thankful that our arrival eased their burden at least a little bit. Before the P-51 their odds of surviving a tour were slim indeed.

On my first four or five missions, there were no enemy fighters to be seen. Who needed the added hazard of a dogfight on top of the misery of weather, flak, and possible collisions we already had anyway?

Even their absence during the next few following missions didn't bother me too much. However, it wasn't too long before the score some pilots were making on other missions, or in other places, began to arouse my old competitive spirit. Where was the enemy? Here I had phenomenal eye sight, was top drawer in aircraft recognition and had plenty of combat hours — but I had yet to see one enemy aircraft. No other pilot around me did either. I was beginning to feel jinxed as my combat hours mounted with still no appearance from the enemy.

Flying so many missions without seeing an enemy aircraft may seem strange, but it wasn't really. Take the record of "Kit" Carson from my squadron, for instance. He would become the group's highest scoring ace with 18½ victories. He volunteered to fly a second tour for a total of 105 missions. All of his kills were accomplished on less than a dozen missions. If you threw in an unlikely ten missions where he might have seen E/A without a kill, that would still leave something like 80 missions that he flew without seeing an enemy airplane. Out of the 59 missions I flew, over 50 were flown without seeing a single E/A.

I hated instrument flying with a passion. If the ceiling was high enough at takeoff, I would assemble the whole flight in tight formation just under the clouds, take one last look to make sure the boys were tucked in tight and immediately start up through the clouds. Strange things happened sometimes. Once, after checking our typical finger formation, wingman on one side, element leader and his wingman on the other, I climbed into the clouds and came out on top and checked the boys again, two on one side — and two on the other? I don't know where, or how, the extra

P-51 joined us in the clouds, but I was sure glad it wasn't by way of collision.

On another day the ceiling was right on the tree tops. Under those conditions, your wingman is already tucked in from the formation takeoff and you just immediately climb on into the soup. on this day, just after I was thoroughly into the clouds, a voice screamed on the radio, "Get out of that thing! get out of that thing! You're on fire!" Who? Who? Thanks a lot fella! After my tongue was back in place, I continued the climb-out. I never did find out what that was all about.

Destiny seemed to enjoy toying with my hatred for flying on instruments. On another day we had a low overcast where the first to take off were reporting on top at around 2400 feet and all clear above. As I waited for my turn to depart, I took comfort in the fact that, at least on this day, my time on the gauges would be relatively short. Soon I took off with my wingman and made an easy circle of the field so my element leader and his wing man could catch up and settle on my other wing. With everyone tucked in close I headed on up, still in an easy circle to insure staying in the near vicinity of the field. As I passed through 2000 feet, I was really looking forward to the clear sky that would soon be showing up. Other flight leaders were still reporting on top at around 2400 feet. At 2350 feet, I was expecting the clouds to thin out at any second. At 2500 feet I was still expecting — and at 2600 — what was going on!? On and on I climbed, 3500 feet — 4500 feet — and others behind me were STILL reporting on top at around 2400 feet! I didn't have the foggiest notion of what was going on but there was nothing I could do but keep climbing and wonder what the hell the other guys were imagining; 6000 feet went by, then 7000, then 8000 feet! This was really getting ridiculous! Finally at 8500 feet I broke out on top and, sure enough, the top of the overcast was at about 2400 feet all around — except right under me. There, standing straight up like a giant ghost, was a two or three mile wide column of clouds that I had somehow managed to stay right in the middle of!

The ratio of kills to losses for the 357th was heavily in our favor, but still, the number of losses from our original charter members was tragically high. My own squadron would suffer a fatality rate of over 25%, grim statistics by any standard. Others went down from mechanical failure, flak, enemy fighters, or "friendly" fire, but they would survive the war as POW's or escape through the underground. More than one would suffer crippling injuries that would stay with them for the rest of their lives.

The subject of the tours raised a sore spot with me. No one had a greater desire to get home than I did. My home sickness for Helen weighed heavily on my mind. When we first started flying combat a tour could be completed in something like 25 missions (if they were long enough) but that didn't last long. It seemed like the powers that be were extending the length of the tour about every two weeks. I never shirked my duty or avoided going on a mission as I really wanted to finish my tour and go home. If they weren't extending the length of the tour before I could finish, then something else would come up.

For instance just when I was getting a pretty good start on my tour, one experienced pilot had to be picked from each squadron to go to Goxhill for a few weeks as an instructor pilot. Goxhill was a training base to give new arrivals from the States about ten final hours of intensive training before being assigned to a fighter group.

Hershal Pascoe and Hubb Hubbard were picked from the other two squadrons. Hubb had been severely traumatized by the loss of his brother Lloyd and hated flying combat with a passion. He always managed to wear the heels of his shoes crooked and joked that it was caused from being dragged out to his airplane to fly.

Posed picture of H. Mace, without hat, giving "cockpit check" to student fighter pilots at Goxhill. Air Force photo

We had lots of time to just sit and talk once the training chores were over. During these sessions I got some of the stories out of Hubb that caused him to win the disfavor of his C.O. He did tend to be a little on the wild side. The worst of his episodes resulted from the kick he got out of scaring unwary pilots back in the old P-39 days. He loved to sneak up on them from above and behind and dive down and pass under them. As soon as his tail cleared their prop, he would pull up in front of them. Besides the sudden appearance of an airplane four or five feet in front of the unwary pilot, the fairly severe buffeting caused by Hubb's propwash could really shake a pilot up — literally. The door frames on the P-39 were right at temple height and any turbulence banged your head from one frame to the other. It was enough to make any pilot cranky.

This trick made Hubb one of the less popular pilots in his squadron, especially as one of his "victims" was one of the flight leaders. Although the visibility out of the P-39 is good, it was still difficult to see directly behind. So Hubb would watch the other plane as far as was practical and then just time his pull-up in front of it. You guessed it. on this occasion, Hubb goofed and lost a fair portion of his tail. Even worse, the flight leader's plane was so badly damaged, he was forced to bail out. With the loss of so much vertical fin Hubb's plane was totally uncontrollable with the gear down and uncontrollable below 200 MPH with the gear up. The sensible thing would have been to bail out, which everyone else in the air was urging him to do. But the thought of bailing out absolutely petrified Hubb and he refused. He wound up belly landing the plane on the hard runway at a little better than 200 MPH! He said sparks flew over the wings like water spray from the bow of a speedboat! Needless to say, that was the end of that form of horse play, and it's my guess that only a brother high in the Pentagon kept it from being the end of Hubb's flying.

Like most of us, his image was important to him. One of the most image enhancing things for Air Force pilots was the "thousand hour hat." Pilots had convinced higher brass that it was necessary to remove the metal ring that kept the standard issue hat smooth and firm to allow the use of a headset while flying. If you managed to have the hat look a little dirty and disheveled to boot, so much the better for looking like a real veteran. Hubb carried this to an extreme. His hat looked like a mechanic had used it while working under a greasy car. But Hubb was a good kid with a terrific sense of humor; his imitation of Gunga Din falling off the tower blowing the warning bugle was a riot.

The pilots arriving at Goxhill were very green, usually with no more than ten hours of flying P-40s. Not only was this minimal experience but they were further handicapped by the restrictions put on them by the sob-sister induced limitations of no buzzing and no acrobatics! Cry-baby mothers couldn't stand to hear of a pilot being hurt or killed in training. They apparently preferred sending their boys off to combat totally unprepared. Our stiff backed C.O. at Goxhill was also a Stateside throw back who wouldn't allow us to fly below 500′. Really! I wonder what rule book these jokers thought the Germans were going to fly combat by!

Glen "Hubb" Hubbard

I'll guarantee you the training I gave these boys in the ten or so hours available was combat oriented. There were a couple of close calls, but I didn't lose any of them. One managed to chew most of the rudder off of another when they were just off my right wing. It didn't damage the prop at all and the pilot in the near rudderless '51 made a safe landing. Even with the intense training I tried to give them, it was a shame to send them off to combat groups with only about 20 hours of flying time in a fighter plane.

One of my students was Frank Gailer who went on to join my own 357th Fighter Group. He shot down five E/A to become an ace but unfortunately was shot down by another P-51 on his last engagement. He survived the ordeal and remained in the service after the war to retire as a General. He loves to needle me at our reunions by claiming that if he had had a better instructor, he wouldn't have been shot down!

Sharing the field with us at Goxhill was a P-38 outfit doing the same job we were. The '38s were old war wearies being cared for by a bunch of disgruntled ex P-47 crew chiefs. The accident rate these inexperienced boys were having in those old dogs was atrocious. It was causing a real morale problem among the pilots. As a result, the powers that be decided to send Tony LeVier, who by this time had become a top flight test pilot for Lockheed, over to Goxhill to show the kids just what a P-38 could do. I watched the demonstration and it was very impressive indeed. Tony LeVier knows how to fly!

I became acquainted with one of the P-38 instructors and he talked me into one of them and ran me through the cockpit checks, starting procedure and all. As soon as I got it running, he patted me on the back and urged me to take it out and fly it. I was very tempted but in view of the dismal record of mechanical failures that these junkers had, I elected to wait until I had the chance to fly a good one. Had I known that the chance would never come, I certainly would have accepted the offer, and there have been many times since that I've regretted not accepting.

British defenders had pretty well put a stop to big air raids by the Germans, but small one or two plane bombing raids were still common. For some reason, I had virtually no fear of these air raids and wasn't about to pass up the chance to watch another side of the real war in person. The raids were always at night and the sky was often full of bursting anti-aircraft fire or the sound of the high revving Mosquito night fighters pursuing the growling twin engine German bombers. Hearing the machine guns fire and watching as the German planes burst into flame and fall to the ground was quite a sight. Even though bombs fell fairly close on occasion, I never failed to stand out in the open to watch, well almost never.

The Germans usually came over before bedtime, but one night they came over long after I was comfortably in bed and in no mood to leave it. Hubb and I shared the same room and the air raids scared him to death. I was just barely aware of the sirens going off and Hubb rushing for the door. He no sooner looked out than he began yelling at me. "Get up Mace, Get up!, you better get up so we can run to the bomb shelter! This is a bad one! Really a bad one! Get up!" I could hear bombers and bursting flak pretty close overhead but I really wasn't that interested in getting out of bed, but Hubb was so insistent that I finally agreed to get up and look for myself. I guess Hubb was too scared to even go to the shelter by himself. I sat up on the edge of the bed and leaned over to slip on my shoes and ambled over to look out the door. "Naw, that doesn't look bad enough to get out in the cold for, I'm going back to bed." I sat back down on the edge of the bed to slip my shoes off and there between my feet was a four inch piece of jagged flak stuck in the floor. I looked up to see a nice round hole in the roof. I had missed by seconds getting that piece of flak in the back of the neck. Oh well, it was nothing that wouldn't wait until morning. I went back to bed.

Another time during a pretty good raid, I was standing alone outside the shack watching. Hubb had long since made it to the shelter. Soon a JU under attack by a Mosquito salvoed his stick of bombs just as he was coming in my direction. The bombs were exploding one right after the other and I could tell by the spacing, and where the last one fell, that I was ground zero for the next one. I briefly wondered if I would hear or feel anything as I blew up. Luckily, I didn't have to find out — another bomb never fell. Some people are just slow learners, I guess, I never gave up watching the air raids.

Luckily the chore at Goxhill lasted only a few weeks, and when I returned to Leiston, Sweet Helen was gone. I was issued a brand new P-51-D that I promptly named Sweet Helen II which I flew for the balance of my tour. Flying the P-51-B had been a real pleasure right from the beginning. It was even better with the Malcolm hood, the clear plastic canopy that could be slid open at slower speeds. It was especially helpful on some landings where things were a little tight and you felt the need for a little more visibility. During a landing with the canopy back one day, I must have completely forgotten that old BT incident. Just as I was about to touch down three point, that fool scarf blew over my face completely blinding me again. I think I left four red streaks down my face from clawing the thing away. This was the day I gave up my lucky scarf!

The P-51-D and the P-51-B each had their points. The "B" was the better flying plane, especially after the "D" had the 65 gallon fuselage tank installed. The "D" was also heavier by two additional .50 caliber machine guns. The six machine guns of the "D" were vertically mounted and were reliable under any condition; whereas, the four slant mounted fifties on the "B" were noted for jamming under the slightest "G" load. The "D", with its tear drop bubble canopy, had just a slight edge in visibility over the Malcolm hood equipped "B". If the guns on the "B" had been reliable, I believe I would have chosen it. The only bad spot in either P-51 was the spot you sat on, the damned unyielding life raft. It was murder. For me, the worst of the pain came in about the 3rd hour. After that, a sort of numbness came over my rear-end and I could hack the balance of the mission.

It wasn't long after my return that we were lined up on the runway getting ready to take off on a mission. The time for my take-off was getting close when Sweet Helen II shook to the sound of her poor tail assembly being chewed by the prop of the plane behind me. I was very agile in those days — and watched the second half of the accident from out beyond the wing tip. I had no way of knowing the prop was going to stop chewing at the vertical fin. Oh well, my airplane was repairable and it gave me the day off. The prop on the offending plane came through almost unblemished.

If you look real close you may be able to see "Sweet Helen II" written on my P-51D directly below the exhaust stacks, as we head out for another long mission.

On one of the few days that permitted such a luxury, a few of us pilots were sitting in front of the Nissen Hut enjoying the rare sunshine. Stories were being bandied about, close calls rehashed, feats of daring revealed, etc. when one of the guys eyed a pair of tall poplar trees a couple of huts over. They were spaced a little closer together than the span of a P-51. "There's a pair of trees no one will fly between." In a fit of mental lapse, I quickly declared that I could do it! I underestimated the other pilot's willingness to challenge and was immediately asked by all to put up or shut up. What had my big mouth gotten me into?

It was surprisingly easy to take your plane up for a little noncombat flight. All you had to do was come up with some lame excuse about some small item that needed checking with a test flight. So it was that I was soon airborne and looking at those two trees from the air. Somehow they looked like they had moved closer together. Well, to maintain my credibility, I just had to give it a go. I came at the trees in the arc of a very tight turn which put me fairly vertical to begin with. Then I quickly went to full knife edge and passed between them. When I pulled up and looked back at the trees I could see that the prop wash had made them wave until their tops almost touched each other. With the feat accomplished I could now go back to the bull session with some satisfaction. However, by the time I got back the bored bull slingers were long gone and there was not one pair of hands left to applaud.

The P-51 proved to be fairly trouble free. In the beginning there was quite a siege of aborts from rough running engines until suitable spark plugs were found. This caused me to abort two or three times in both my airplanes. It was not that the roughness was so bad that total failure was imminent, but no one should be asked to head out on the kinds of missions we flew with less than a good running engine. Virtually everyone in the group would experience this problem at one time or another. On one mission as many as 15 planes had to abort.

There were occasional, although rare, total engine failures. Having the only engine you have suddenly come up totally silent is a heart stopper under any circumstance. But, if it's pilot error and at a particularly inopportune time, as happened to me once — thank goodness for being in a single-seat airplane!

It happened when I was returning from a mission over the Peenemunde area of Germany using the usual route which required crossing the Danish Peninsula, the flak infested island of Heligoland, and flying over the quite a few miles of hostile North Sea back to England. My good old group had become totally separated one more time and I was returning alone. Once comfortably past Heligoland I felt like relaxing and decided to buzz the rest of the way home. I dropped down between the chops of the North Sea and enjoyed the rush of skimming the ocean. Buzzing between or over the larger peaks of greenish blue sea was a nice refreshing change of pace but, with no land in sight and still far from home, I needed to carefully monitor fuel consumption. When the right time came I switched to my last full tank. I made some quick calculations and decided I'd feel better if I switched back to the previous tank and use just a little more of the 15 or so gallons left in it. It had been a long mission. I went back to playing tag with the white caps and subconsciously must have thought my fuel problems were over. My reveries were soon rudely interrupted by sudden and complete silence from the engine. Most startling when you're staring at North Sea white caps right in the face. Very quick corrective actions saved the day, however, and only lonely embarrassment remained.

Total mechanical engine failure does not leave pilots with too much to choose from. I was watching the day's activity once when my good buddy K. Vogel was about 5,000' overhead on a test flight. His engine quit cold and in the attempt to make the field, he fell short and totaled the P-51 with little physical harm to himself. The effect it had on his confidence was something else, however. He was so up-tight on his following three flights that each landing resulted in relatively minor accidents. The worst was a nose-up that knocked the prop and gear box off the engine. In bracing himself, he inadvertently pushed full throttle. Without the prop, the engine wound up to scream in a manner that I had never heard before or since. He ultimately overcame his problem only to be killed in action. It was a great loss.

Two good photos of my A/C, showing the different camouflage used.

On another occasion, I was about to beat up a competing group's field and was looking down at one of their P-51s approaching to land just as his engine failed. He too, fell short and his crash was about the same as Vogel's. I watched as wings and things came off, but the cockpit area remained reasonably whole so I assumed the pilot was OK. In order not to interfere with rescue efforts, I abandoned my planned mock attack and went on my way. Of course, there were some who just never returned from missions and in the absence of witnesses, the cause of their loss will never be known. No doubt engine failure could be blamed for some.

A more rare cause for an abort was such as I experienced deep in Germany once. Fuel was always used out of the drop tanks first, so no monitoring of fuel gauges was necessary until internal tanks were put to use. Deep into the mission, I switched to internal fuel and a glance at the fuel gauges showed zero on one of the main tanks. The fuel had siphoned or leaked out unnoticed somehow, which left me with insufficient fuel for the balance of the mission. I had to return to base. Normal practice in individual aborts is to have one additional plane accompany the troubled craft on its return through hostile territory. I hated to deprive the bombers of any more fighter cover than necessary so, since I was a flight leader, I elected to return alone.

Since the accuracy of enemy flak fell off heavily above 30,000′, and the likelihood of being bounced from above by E/A at the higher altitude was less, I elected to fly at 31,000′ on my return. A single high flying fighter doesn't draw as much attention from the enemy so I had settled down to a rather quiet return flight. I still kept a sharp look out, however. About half way home, a tremendous boom beneath me penetrated the '51's noises and Sweet Helen II was thrown almost inverted. Flak! My ho-hum opinion of the German's high altitude accuracy was rudely shaken. It was the nearest miss I had experienced. A little dodging shook them off, and periodic course changes for the balance of the trip home prevented any further surprises. Amazingly there wasn't a single hole in my airplane. A surprise, since I had earlier taken a hit in my right fuel tank from a burst I couldn't even hear.

One other abort I had was caused by a coolant shutter motor burning out in the open position. This caused the engine temperature to stay too cold to run properly, forcing a return to base. On climb out through 8,000′ of overcast on another day, I gave my usual unshakable attention to the blind flying instruments. When I broke out on top, I immediately scanned the engine instruments. The oil temperature needle was clear off the gauge and crossing hands with the fuel pressure needle. My first thought was that it was broken. Just to make sure, I stared at it briefly and in a few seconds detected more movement. It was not broken, and the engine was dangerously hot! I throttled back, laid the P-51 over in a Split "S" and penetrated the clouds in a dive for home. The oil radiator shutter motor had burned out, this time in the closed position. For the amount of flying done, these few discrepancies were relatively minor, however. The overall reliability of the P-51 and the maintenance people were very good. My crew chief, Ray Smith was one of the best.

Line mechanics worked long hours in all weather to keep 'em flyin'. Ray Smith was my crew chief for Sweet Helen II.

Occasionally we would have a little easier mission that was referred to as a "milk run." They were somewhat shorter than usual and perhaps the target might be a little less heavily defended. This did not mean you were not likely to get shot at. You were bound to be a target of flak every time you entered enemy territory, at least to some degree. For the most part, our missions were long and difficult. As bad as the 25% fatality rate for the 362nd

My good friend, Harry Ankeny, smiling as only a combat pilot who has completed his tour can. August 16, 1944

charter members was, it pales alongside the 60% suffered by the 363rd Squadron. Our original members exacted a heavy toll of enemy fighters but only by paying an awful price. Flak deserved respect. We had one C.O. for a while who transferred in from outside. He had an absolutely scathing opinion of flak. "Don't worry about flak, it never hits you, just ignore it!" We had always dodged. It was only a short time until he didn't return — downed by flak.

If you were to interview 20 different pilots who had all been involved in the same war-time action, you might think that you were hearing accounts from 20 different actions. This doesn't mean that any of them are lying or intentionally distorting the truth. It's just that what's important enough for one to remember, may mean nothing to another, or one may have seen or done things that the other did not. I have found in researching this book that some things that I assumed were facts proved to be wrong simply from a lack of all of the information. Sometimes, after too many years have gone by, a person's memory can blank out a day entirely. A good example was brought home to me while reading the diary that Harry

Major Irwin Dregne's C5-Q, an early P-51D (notice the absence of a fin-strake) and another 364th Squadron A/C, C5-U, a P-51C with a bulged Malcolm canopy.

Ankeny kept of his combat tour and was kind enough to let me use for researching this book. Quoting from the diary date of 29 June 44:

"Capt. Williams came in our barracks at 9:00 P.M. and wanted three of us to fly an Air-Sea Rescue mission. Lt. Mace, Wallen and myself volunteered and went out for about an hour and half, landing just after dark. We took off all together, 3 abreast and looked good. A large Air-Sea Rescue plane was being towed back to the coast after being forced down in the North Sea. We just flew around it and covered it and the tow boat. There were also 3 P-47s covering it for a while. Not much of a mission and got 1:25 hours combat time."

The hook here is that the entire episode has completely faded from my memory and yet a check of my flight record does show three missions flown that day. Those English summer days were long ones.

Sometime in mid summer, the big wheel mission planners decided that running shuttle missions using Russia as a temporary base was a keen idea. The first effort out of England was something less than successful. Most of the bombers were destroyed while sitting on a Russian air base during the night. A reconnaissance plane had spotted the juicy target for a German air raid. Our fighter group was picked to cover the second shuttle mission. Since it would be escorted by our group alone, the decision was made to compose a fourth squadron out of the spare planes and pilots from our three regular squadrons. It made an impressive fighter group. Again quoting from Harry Ankeny's diary, he had this to say of the shuttle mission.

"RUSSIA-ITALY-ENGLAND SHUTTLE MISSION, 6 Aug. 1944. Today was Part 1 of our long shuttle mission to Russia, the first leg of our trip.Since we were the only group to be escorting the bombers thru the targets, we formed a group made up of the maximum number of planes. On all previous missions we used three squadrons but this time we formed a fourth, called the "Victory" squadron made up of 11 flights from all three squadrons. Major Dregne led us while Col. Graham, our group commanding officer, led the group. My job was leading green flight in "Victory" squadron with Lt. Sehl on my wing and two 364th boys comprising the rest of my flight. On the ground I was also to act as our squadron's adjutant.

The bombers took off quite early while we started around 8:00 A.M. After the group was

Gunther Schack, center, a 195 victory German ace who was shot down by Bud Nowlin, left, of the 357th on his return mission from Russia. Gunther was an honored guest at the 357th reunion in 1991. Group historian Merle Olmsted is on the right. Below is a picture of the plane Bud used on that misson on a later date - crashed by another pilot.

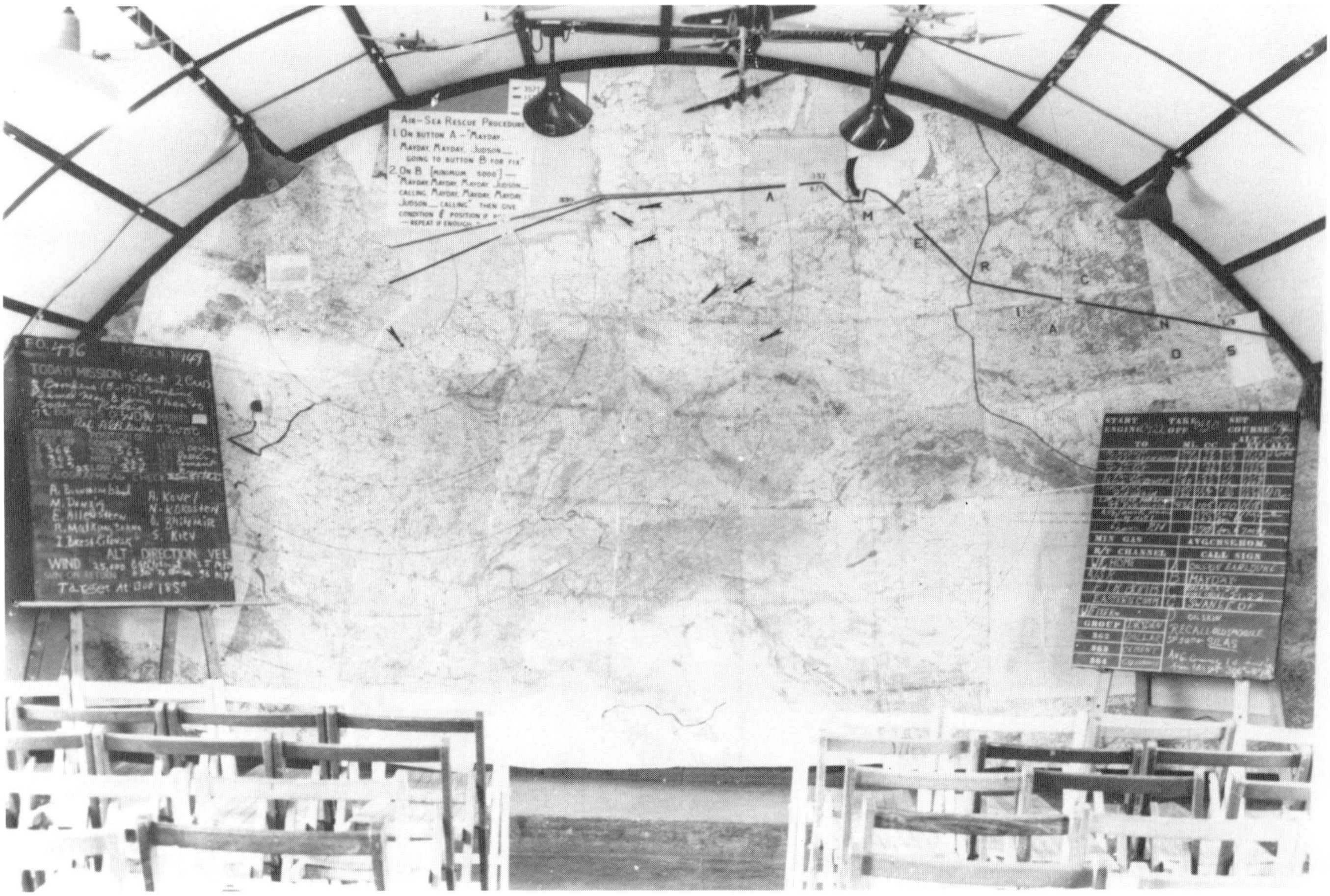

The briefing room at Leisten airfield on August 6, 1944, the day the 357th flew it's first shuttle mission to Russia. Note the mission tracking tape running from England, on the left, all the way across Europe and off the map at right!

formed we headed up along the Fresian Islands and crossed into Germany just below Heligoland and crossed out again between Kiel and Lubeck, Germany. Then out over the water again until we sighted the bombers just this side of Danzig. We joined them and they bombed an airfield just north of the East Prussian capital city and then started for their Russia base. We stuck right by their sides all the way to Russia. Nearing the Russo-German lines we did see a little enemy activity but they didn't stick around long when they saw the bombers were protected by a good fleet of Mustangs. No aircraft was shot due to enemy action. When we were fairly certain we were within the Russian held territory, we let down to about 5000 feet and enjoyed sight-seeing the remainder of the trip. The land looked very fertile but the towns were a bit torn up. We could see smoke far over to our right from Warsaw, which was the extent of the lines at that time. We flew right on into the Ukraine over Kiev and on to our respective fields, the bombers going on to Poltava, while we landed at Piryatin, a small village just 100 miles east of Kiev.

Part 1 took just 7 hours and 30 minutes and we were all plenty tired when we landed. The American soldiers at the field were sure glad to see us and gave us a real welcome. Never saw so many flares shot in the air before. We immediately ate supper and then half of us were to fly down to a field near the Black Sea so our airplanes would be dispersed. At the last minute they decided against it because of bad weather. We stayed up in the hospital area about a mile from the field. I never saw a field so well protected with AA guns. In the early part of the evening, the Russians started warming up the guns and it looked like the 4th of July around the field for an hour or so. The Russians guarding the field looked very rugged but a little old. I imagine all the young blood was up on the front lines. To each of our planes was assigned a U.S. Army Crew Chief and a very young Russian mechanic. Sure was good sleeping that night."

Johnny Pugh on the the wing of "Geronimo" at Piryatin Russia.

After passing the flak saturated island of Heligoland, we were crossing the Danish Peninsula, when I saw a flight of nine FW-190s in a very tight echelon, show style, formation — very pretty! They trooped right by our whole string of bombers and fighters, oblivious to all. Apparently none of our boys saw them either, as there was no mention of them on the RT. They were passing about 1500′ lower and to my right, going in the opposite direction. I found the easy targets hard to pass up. I even made a tentative push over to attack but had to remind myself of the strict orders to leave E/A alone that were not threatening the bombers. All I could do was straighten out and watch them pass out of sight to our rear. After the bombers hit the target, we proceeded to the desolation of Russia where we stayed a few days. While there, we flew a mission over the Eastern Front. Other than flak, the only enemy activity was provided by a lone ME-109. He apparently was looking over the bombers who were below us and to our right.

The set up was a copy of the situation with the 190s over the peninsula. Only in this case he posed a possible threat to the bombers. He passed our whole gang of fighters slightly lower and to our right without seeing us, or our guys him, apparently. I started to nose down for a head on attack but a flight well ahead of me finally saw him and swung sharply around to pursue him from behind. Even though they were well behind him, they had the priority position so I backed off and just watched as they passed under me and disappeared. I have no way of knowing whether they were successful in their pursuit or not.

The target for the bombers on the leg from Russia to Italy was Ploesti. Fortunately it proved to be a milk run compared to the famous, but disastrous, B-24 raid earlier and we flew on and landed successfully in Italy.

What a contrast the weather in Italy was compared to England and Russia. The Americans stationed there were as brown as Indians. We pale faces stood out like albinos in Africa. We stayed a few days and our group flew an escort mission over Yugoslavia while C-47s carried out supply and rescue efforts. It was a limited milk run mission where not all of us were needed so I was even able to spend a few hours on the beach. It was very much like being in Southern California. I can't say much for the sanitation practices in the nearby town, however, it was appalling.

The target for the last leg was in Southern France. As we crossed the Corsica side of the Italian coast, very close to the front lines, I looked down from our position of about 25,000′ and saw the flashes of a sudden artillery barrage. I felt sorry for the poor guys who were the target of all that firepower and wondered who they were. I didn't have long to wonder — big ugly blotches of flak blossomed all around us. We were the target. So much for my mis-directed sympathy. At least none of us were hit and the usual flak over the target in France also missed. The rest of the trip home was routine.

The group was performing beyond expectations and gaining quite a reputation. Not all of it was favorable. Our effective defense of the bombers, combined with some pretty vicious strafing, was getting under the skin of German High Command. Their feelings were made known to us through their propaganda tool, Lord Haw-Haw, in his broadcasts over the radio. in an effort to intimidate our efforts, the threat was made to execute the next 357th pilot brought down alive. The next pilot to go down was Fletcher Adams. It is not known whether he was really executed or died from his wounds but, whatever the cause, he did not survive. In any case the attempt at intimidation was a total failure, as the fury dealt out by the 357th for the balance of the war would attest.

The war took most of my attention in England. Very little of my time went to things other than flying. There were occasional card games and two or three trips to a local antique shop and that was about it. I left all of the pub crawling and party going to the others.

Before Tom Beemer was shot down by a B-17 gunner he did make a discovery that led to a little extra diversion for me. Besides being a handsome devil he was quite an artist. He discovered that the walls of our pilot's room were covered with thick panels of cardboard that had a light colored surface over a darker center and by carefully scribing through the outer layer with a razor blade, and peeling it away, he could create a two tone portrait. He then proceeded to sketch out a beautiful nude girl (What else?!) and carved out a beautiful portrait. Being very happily married my mind ran more towards airplanes so I carved a flight of P-51s peeling off on the panel next to his. I also started to keep the score of kills for the squadron by carving a cross for each one around the P-51s. The mounting score soon far surpassed the room on the panel so the tallying had to be continued on the end wall of the pilot's room. I've often wondered what ever happened to those two panels. The buildings are long gone and I suppose they are too.

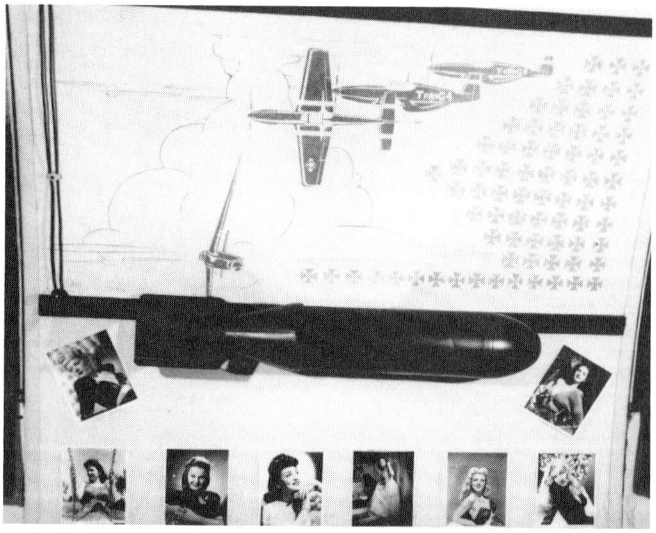

A photo of the picture I carved in the panel of the pilot's room. The original panel was approximately 3.5′x5′.

Being near last in seniority (of our charter member gang) had rightfully relegated me to flying wing for awhile. But war and combat being what it is, it was inevitable that some of those with more seniority than I would ultimately leave the Squadron. All too many would die in combat and many would finish the war as prisoners of war. A lucky few would finish their tours and be returned to the ZI. This resulted in rapid rises in rank and position for those remaining in combat. A member of one of the other squadrons went from Lieutenant to Major in six weeks! For myself I went from being a tail end charley Second Lieutenant to a flight leader Captain in fairly short order.

Capt. H. Mace with group of new pilots at Leiston.

The rapid rises in rank and position also meant a rapid rise in responsibilities, which fell on shoulders already heavy with the hazards of flying combat, home sickness, and the possibility of never seeing loved ones again. Among the responsibilities was deciding who would fly on the next mission and who wouldn't. With very few exceptions the pilots answered the call willingly every time. There were the occasional head colds or other minor illnesses that had to be accommodated. New pilots had to be babied into the flight as much as time would allow. I usually sent them up for a local flight or two, whenever possible, to help get their "sea legs" before being exposed to the rigors of combat flying. This resulted in a rather strange non-combat accident that involved a new replacement pilot to my flight, Lt. Richard Ivan Potter.

I sent him up on a local orientation training flight as tail end charley of a flight of four. When the flight returned there were only three planes. "Rip", as he was known to his friends, had crashed. The nickname "Rip" came from his initials. None of the others knew what happened but word soon came in that Rip had bailed out and was in the hospital.

As soon as he was well enough for visitors. I went to the hospital and walked up to a patient who was totally unrecognizable. His head was swollen to where it looked like a round ball with a nose, eyes and ears attached to it. But, he was completely alert, in good spirits, and tremendously glad to see me.

It seems that when they were well into the flight the leader peeled off from an echelon formation to hit the deck. As often happens in rat race situations, each following pilot is prone to peel off a little higher and later than the one ahead of him and Rip was a little later than usual. They were in a very steep dive when Rip made the mistake of pushing on a lot of throttle to catch up. You just shouldn't do that in a Mustang, especially in a P51-B with a Malcolm Hood. Rip had somehow gotten this far in his flying career without having heard about the sound barrier and its warning symptoms of control stiffening and buffeting which could tear a P-51 apart.

At any rate he said that when the buffeting came on, he thought he had lost the tail of his airplane. By using the virtues of the Malcolm Hood he was able to turn around and look back and found that it was still there — briefly — it came off as he looked at it! The plane then snapped into a violent outside snap roll that threw his head right through the very thick plastic of the hood. The jagged edges then locked around his neck and prevented him from pulling his head back in. At this point he admitted to a (very understandable) moment of hysteria — but soon realized it was getting him no where and he settled down. He then decided to jettison the canopy, a move that I'm certain would have torn his head off! Luckily for him his right shoulder had been dislocated which prevented him from using his right hand and his harness prevented him from reaching the canopy release with his left hand.

My flight beats up our base. Note maintenance hangers in background.

Next he tried to release his safety harness in order to reach the canopy salvo lever. He said the last thing he remembered in the cockpit was reaching for the harness release with his left hand. When he regained consciousness he was aware of being on his back in peaceful, quiet, white surroundings and he thought he was dead. A very brisk wind blowing soon made him realize that he was alive and the white surroundings were clouds through which he was falling at a very rapid rate. Now alert, he groped for the rip cord of his parachute with his left hand and pulled it. He said the 'chute only made one swing before he hit the ground. He was a very lucky boy. It would seem that the only logical answer for his survival would be that he did release the harness which allowed the gyrating airplane to throw him the rest of the way through the canopy and clear of the airplane.

The losses of allied aircraft were very high during the war and the damage to those that were lost ranged from total disintegration to belly landings that left some airplanes reasonably intact. It was inevitable that the Germans were able to gather enough good parts to assemble a few flyable examples of our aircraft. It not only gave them the opportunity to steal ideas and evaluate the performance of the planes but also to put an occasional one to more sinister use. As a result there were reports of captured allied aircraft being flown in action against us. A good example was the occasional report of an unmarked B-17 off to one side of our bomber streams. An obvious benefit of having this German flown B-17 pacing the bombers would be to act as an air borne artillery spotter, as it were.

There is no doubt in my mind that I was involved with one of these captured fighter planes. It happened while I was leading White (#4) Flight on a mission near Berlin. We were escorting the bombers and except for occasional flak, it was going reasonably uneventfully. But pretty soon a lone P-47 slid in from the right and joined up with us. It was not uncommon for lone fighters to join up with other groups when they became separated from their own, so I thought nothing of it. I was a little surprised to see a P-47 this deep in enemy territory, however. He settled down about 400′ to the right and slightly behind me.

I always kept track of everyone around, so when he slowly started slipping behind me, I was aware of it immediately. I didn't like anyone on my tail, friend or foe, especially with eight .50 caliber machine guns armed and ready. I quickly slipped to the right to put him on my left. After settling there for a couple of minutes, he once more started sliding in behind me and I repeated my quick maneuver to the left and began getting a little nettled. When he started a third slide towards my rear, I had had enough so I made a heated pass right at his canopy that clearly said "Stay off my tail!" He got the message and after a short grace period slipped out of my view to the right. Just a few minutes later a voice over the radio yelled: "Look at that P-47, he's shooting at a P-51! Look at him shooting at that P-51!" I never discovered what the outcome was, but I knew we had been the original target.

I had flown nearly 100 combat hours before I saw my first enemy aircraft that I might have been able to engage. It was on a mission to Mersberg, near Berlin. We were escorting the bombers and considerable confusion resulted when a pilot near the left rear reported seeing a couple of bogeys. Surprisingly, the group leader responded by ordering us to drop our external tanks and swung the whole group over and back. I'm not sure why he did, one flight could have handled the situation quite nicely.

For the only time during my tour, my drop tanks wouldn't drop. Neither pushing the release button, yanking the salvo lever, or working them both together would rid me of the tanks. I would have to do both of the above and fire the guns at the same time to shake them off. This required waiting for the skies in front of me to clear before I could fire the guns. By the time I was finally able to get rid of the tanks, everybody else was out of sight, including my element leader and his wingman. I should have had some firm words with him back at base, but the 357th had always been a little loose about hanging together. All of the confusion left just me and my wingman as sole escort for quite a gang of B-17s. my wingman on this mission was an RAF officer learning the American style of long range daylight escort. I believe his name was Wooley, and unlike my RAF instructor in Advanced Training, he was an excellent pilot.

A little while later I was weaving back and forth over the bombers when a little flash of light far down to the left caught my eye. It must have been six miles or more away. With a hard look, I was barely able to make out a flashing silver B-17. Falling B-17s were a common sight, but this one wasn't falling, he was dog-fighting! But with what? More real hard searching finally rewarded me. I spotted a little speck of a plane pulling up after having obviously just made a bounce on the bomber. He was pulling up in a long climb to be out of gun range to make his wing-over and head down for another pass.

I immediately peeled off and headed down while taking one last look to be sure there was nothing around to bother the bombers we were leaving. My wingman was sliding into the best spot there is for staying with a gyrating leader. That spot is in line, slightly lower and with his nose almost under my tail. It is very difficult to lose someone in that position, but they are also next to impossible to keep track of. I didn't even try.

When you point the nose of a P-51 down under power, it will really move out. Unfortunately, when the speck of a fighter made his wing-over to bounce the bomber again I was still so far away that he would be in firing position before I could bring my guns to bear. But, in order to try and save that bomber crew, I raised the nose of my ship and fired a short burst. I hoped the tracers would drop somewhere across the nose of the German so he could plainly see them. I immediately dropped the nose back down and watched as the tracers arced over and down. When they reached the point where they should have been right in front of his nose I was rewarded with the sight of the pilot breaking off the attack and making a run for the layer of clouds below.

I was close enough now to positively identify the E/A as an ME-109 and I very much wanted his scalp. Unfortunately, it was quite apparent that his chances of reaching the cloud cover before I could reach him were excellent so I really let it all hang out. I pushed a little too hard, though, and soon got the stiffening of controls and buffeting that signal the approach of compressibility. Having eyes only for the 109, I never looked at my instruments but my wingman later reported that he quit looking at his as the airspeed indicator went past 550 MPH indicated; he had seen enough! A P-51 is quite capable of tearing itself apart at too fast a speed so it was imperative that I slow things down a bit. This was the final assurance that the German would make the clouds before I could close to shooting range. I gave a short burst from out of range anyway — with no effect other than to relieve my frustration. I was not giving up yet, however. I bored on down and leveled off over the clouds to about where he should be. Viola! There he was! He had leveled off too close to the surface and although he couldn't see horizontally, I could clearly see him. I could even see that his eyes were glued to his instruments and he seemed to be as tense about instrument flying as we were.

I knew that sooner or later he would feel safe enough to either come up or go down. I just waited him out. It wasn't long before he started descending. This was no time for finesse; throwing caution to the winds, I just jammed the stick forward and dove. I wanted to be below, waiting, when he came out of the clouds. fortunately, there was adequate ceiling under the clouds and I was below them first. Soon the dark figure of a fighter descended out of the clouds right in front of me. I had him! But wait, it was very dark under the clouds and I was partly snow blinded by the white clouds above and had lost track of my wingman. Had the ME-109 veered off and was this my wingman trying to find me? If I was going to shoot at someone with six .50 caliber machine guns, I wanted to make damn sure it was the enemy. I risked being detected by easing forward and to the right enough to not only make out the unmistakable form of the ME-109 but the German markings as well. I slipped back behind him and put my stupid gun sight's pip right on his rudder. One might wonder what a person feels when he is about to pull the trigger on another human being. I felt nothing. There was no apprehension, fear, guilt, or excitement. It was just a job I had been trained to do. The enemy fighter was no more than a target that had just tried to kill my fellow airmen. I pulled the trigger.

Damn that sight, it was shooting 20′ too low! I carefully raised the pip to 20′ over his head and fired a good burst. All those .50 caliber shells hit home and I couldn't believe the destruction. So much trash and smoke came off him that I could no longer see him. Fearing that I might run into him if he were dead in the water, I throttled back and pulled up. It was a good decision because I slid right over the top of him. In order to keep him in sight, I had to roll over on my back and look up — down. The pilot was still alive and looked up at me. For a moment we were eyeball to eyeball. He was very busy shedding his safety belts preparing to bail out. I quickly rolled to the side and came back in, hoping to kill him before he left his plane. My effort to bring my guns to bear again were too late however, he had wasted no time in bailing out.

After a brief glance at his drifting parachute and noting my lowering fuel level, I decided that it was time to head for home. I had finally downed my first enemy aircraft. I could have killed the German in his 'chute, but shooting airmen in their parachutes was an unwritten no, no. We probably had ten men bail out for every one of theirs. Why try to kill the pilot? I had no qualms about it at all. I knew perfectly well that if he lived, he would be up in another fighter the next day doing his best to kill fellow allies again. One captured German pilot came through our base that had been shot down four times. His score was eight allied airplanes and most of them were kills that he made after he had been shot down the first time.

I was completely alone as I climbed for altitude on the return trip home. It was somewhat hazy, but soon another P-51 came into sight on my left. Since it's not too smart to be alone in hostile territory if you don't have to, I decided to join up with him, whoever he was. It turned out to be good old Harry Ankeny from my own squadron.

Being real eager to tell someone of my first score, I used sign language to do my bragging to Harry and it brought back an enthusiastic O.K. This victory would be the first of three during my tour, although I claimed only two. They were the only shooting contacts with E/A I would have. There were three or four other missions where I saw enemy aircraft but they would be under the guns of others or untouchable for other reasons.

Weather was a continuing problem. On the way to one of our targets in poor visibility, I saw one of the most awesome explosions in the sky I had ever seen. It left a great mushroom cloud and pieces trailing smoke in every direction. It gave the appearance of a huge monstrous octopus. I was told by intelligence that a B-17 with mechanical problems had turned back and hit the leader of a three-plane element head on. All four planes with fuel and bombs had gone up in smoke with 40 young Americans.

Poor visibility could also cause anxious moments in the strangest ways. As an example, I was returning from a mission alone when the visibility was particularly poor. It was always necessary to keep a sharp lookout, but under the prevailing conditions it was especially important. As I was scanning the skies, I looked straight overhead and was surprised to find myself about 1,000′ below a box of B-17s. I was even more surprised to see that they had just dropped their bombs! The revelation demanded some immediate action: I moved from that bit of sky to a more favorable one with amazing speed.

A big operation was planned that was supposed to grab big chunks of enemy territory in Holland. It was code named Market Garden. The Garden part of the operation was carried out by airborne troops and the Market portion was the escort and cover done by fighters. The 357th played a heavy part in the Market operation. The book and movie titled "A Bridge Too Far" was based on the Garden portion of the operation.

On the 18th of September, 1944, we had successfully escorted a large mass of C-47s carrying paratroopers to Arnhem and had climbed to about 15,000′ to patrol the area. We began flying a heading that put the sun nine o'clock high to us. I kept glancing in the sun's direction since it's a favorite spot for enemy aircraft to launch an attack from. Soon I spotted a speck near the edge of the sun. It was no immediate threat to us even if it was an E/A, so I delayed saying anything until I had a better look. After a little change in position, I was able to put my fingers over the sun and see a whole bunch of specks! They weren't in any discernible formation and weren't acting like our side. I had to assume that they might be the enemy. It's impossible to count numbers under these conditions but I knew what our group of nearly 50 looked like and the group of bogeys looked to be at least three times that size.

Since they looked very suspicious, I called them in to our C.O. and he responded with something like, "Well, let's go up and take a look." With that he immediately swung the group into a left climbing turn. This move put me dead last since I had my flight on the far right of the entire group. Even being a flight leader didn't seem to keep me from being haunted with tail end charley positions. After we had climbed a while, one of the specks detached itself from the main group and came diving down 500′ above and to the right of us.

It was an ME-109. Now there was no doubt that the main group were no longer just bogies, they were bandits! I marveled at the discipline of the group for not breaking up and chasing the bait. Obviously the important thing was to attack the main group and shoot down as many as possible. It would have been an easy cop-out for someone to break off and chase the lone ME-109. After all, we were climbing up to face a vastly superior force that had altitude and the sun in their favor.

Chuck Weaver, my wingman (left), and I relaxing on the water.

The Germans had ideas of their own, however. They obviously wanted to break us up and sent down fighters in twos and threes to attack. The first two seemed to come down just spraying at the group in general. Two or three of our boys took them on while the rest of us kept on climbing. The first two were followed in quick succession by a string of twos and threes coming down and firing away. The ones in our group being singled out had to return the attack of course.

In short order the whole group was involved in one huge dog fight with planes rolling, diving, and looping with some going down in flames. For a moment the scene looked exactly like the depiction of World War I dogfights as shown on the cover of pulp magazines that I avidly read as a boy. I and my wingman, Chuck Weaver, were the last to be still climbing towards the main bunch of E/A. My element leader and his wingman had deserted me sometime earlier. There appeared to be about 100 E/A remaining in the main bunch and just what I proposed to do with them once I reached them was yet to be determined.

The question soon became academic. An ME-109 dropped down to my level just out of gun range in plain sight — bait! I guess I wasn't supposed to see the two others that settled down about 2,000′ over my head. They were two shooters that were banking steeply from side to side waiting to see just when I went after the bait which was weaving tantalizingly back and forth.

I refused to cooperate with their plan and continued climbing, only now it had to be after the two shooters. I could no longer monitor the main batch of E/A and never saw them again. I had my hands full going for the two shooters and keeping an eye on the bait. At this turn of events, the bait elected to "S" turn ever closer and it was soon evident that he was becoming a threat. His next crossing of my nose would put him comfortably within shooting range so I decided to quickly drop my nose at the right moment and dispatch him with a quick shot and resume my climb at the two shooters before they had a chance to react. The right moment came and I dropped the nose, fired, and resumed the climb before the bullets even got to him. But, drat! I'd forgotten about that lousy gunsight and it was a clean miss. All it accomplished was to irritate the hell out of the 109 pilot and he immediately broke into a vicious attack. Maybe I didn't know that gunsight but I did know dogfighting. I immediately countered and gained the upper hand even though the 109 pilot was good, very good. I had several more high deflection shots that missed for the most part. Through it all I still managed to monitor the two shooters above and they went from appearing somewhat

confused and tentative at the turn of events, to looking more aggressive and about to come down. I was alone now, my wingman having spun out trying to stay with me, and I felt something drastic would have to be employed to get rid of my antagonists.

I knew from witnessing two previous accidents that the Mustang's propeller could chew the tail off a fighter with little damage to the prop. Since I hadn't been able to hit him with more than one or two isolated rounds, I decided to go for a collision with all guns firing and take a piece of his tail off with my prop. When contact seemed imminent I ducked behind the engine in case I bit off more than I could handle and waited — no collision. My day was not going well. However, either more of my bullets found home or the German had just had enough. He bailed out and I immediately re-established my climb for the two above who once more became quite tentative. At least now the odds were getting better.

As I climbed, I kept scanning the skies to be sure I wouldn't get surprised by someone else. Soon a quick glance behind revealed two fighters coming up on my tail. I didn't think it was possible that the two 109s above could have pulled that off, but when I looked up they were gone. The two on my tail seemed a little too far back to be them, but when someone is attacking you, you don't ask too many questions. A near stall climbing speed is not good for maneuvering. I needed a lot more than that so I put the nose down and poured the coal on. It took a lot of self control to keep straight ahead and wait for the right time to break. Too soon and they would just turn inside, too late and you were a perforated target. I didn't actually look at them, just quick glances out of the corner of my eye to check their distance. When the right time came I made a head sagging break to the left. On the radio I could hear, "Where'd he go, where'd he go, was he a 109?" By then I was on the tail of two greenhorns from my own squadron, so I came back with, "You're lucky I'm not a 109, I'm on your tail — and you'd better join up with me. " I can only assume the 109s above, who were already nervous, saw the two P-51s coming and decided to split. Why the greenhorns thought

I was a 109 and didn't see the two I was after, I'll never know.

By then the skies were absolutely clear and fuel and ammo were low. It was time to go home. The group had another one sided victory. I would guess that close to 50 of the enemy aircraft out of the bunch had been engaged. Twenty-six 109s and 190s were destroyed with the loss of two of our own. Not bad when initially we were outnumbered about three to one with the enemy having the advantage of sun and altitude. A surprising twist to the scoring was that none of the group's high scorers made kills on this mission. All of the victories were made by pilots that had no more than one or two kills to their credit.

It would be difficult to recall a dog fight where any part of the group came out as the loser. When it came to counting losses for either side, the boys of the 357th were invariably the winners of the numbers game. Even when it might look as though an engagement had turned out tragically, as happened to three pilots of one of the other squadrons, the numbers count still made them come out winners. They had become separated on the climb out through weather and were never able to find the group. They continued on anyway and ran into a large force of E/A. In the engagement that followed, all three were shot down, but not before they had shot down five of the enemy. Fortunately all three survived the war and were able to tell their stories.

In time I became Operations Officer of the squadron which is second in command of the flying end of things. This required me to lead the squadron whenever it was time for the C.O. to stand down. My days of flying tail end charley were well behind me!

While I was leading the squadron on one of these missions, we were over a layer of overcast that completely covered the target. I just happened to bring the squadron alongside the bombers just as they reached the target. This is always the center of the heaviest anti-aircraft fire and because of the overcast the Germans were firing by radar. Surprisingly the big black track of flak was exploding halfway between my squadron and the bombers. I wondered if it were possible that I was close enough to the bombers that we made one big blip and they were hitting the middle of it? It seemed possible. I decided that if the opportunity presented itself, I would try this trick again. It would be difficult since the target couldn't be seen. All I could do was estimate the distance from the I.P. (Initial Point, where the bombers made their last turn towards the target). A few days later I had the opportunity to try and duplicate the feat. Unfortunately, I wasn't able to properly position the squadron in time. As the bombers came under fire over the target, one B-17 exploded, another went down in pieces, and another slid out of formation badly damaged. Win some lose some. It would be difficult, indeed, to overcompensate those combat bomber crews for what they had to endure.

362nd FS beer party. The three men in front are all aces: John England with cigar, Len Carson holding cup and Joe Broadhead in front.

Leaving for one more. I still have the map I'm holding in my left hand!

On another mission deep in Germany the anti-aircraft fire was extremely thick and inflicting heavy damage to some of the bombers. They had no alternative but to plow straight ahead over the target. We were under fire too but at least in our fighters we had the option of twisting and turning or whatever it takes to evade the flak. Of course, as often happens, it scattered us to the four winds and my wing man and I wound up separated from the gang I had been leading.

As the bombers and scattered escort were leaving the target area for home, I saw the smoke trail of a rapidly climbing ME-163 on the Russian side of the target. The tactics usually used by these rocket planes was to climb nearly straight up to about 40,000′ or 50,000′ and then nose over for a gliding diving attack on the bombers. They were generally ineffective as defensive weapons and this one was no threat to the bombers. I decided to head in his direction anyway, hoping that I could find him on his way down and make a kill. After the smoke trail ended at the top of his climb the plane itself was not visible, but I made a search of the area I thought he would be descending in anyway. It was a fruitless effort.

I headed back towards the now disappearing bombers and fighters — and right into the middle of a horrendous flak attack! I had never seen such accurate big bore flak. It followed every twisting, diving, climbing maneuver I could muster. It was booming in my ears and jarring my plane. It took about 15 minutes of frantic action to keep from being hit and finally break into the clear. By this time the bombers and escort were long gone and I headed for home well behind them. I soon spotted a lone speck low and ahead and dove for it. It quickly materialized as a crippled B-17 straggler with one gear leg hanging down, one engine feathered and who knows what other damage. These boys were indeed in an extremely helpless situation. The crew was not expecting P-51s to be coming up from the rear so the pilot dropped the other gear leg in the accepted manner of surrender. This procedure was accepted by the Germans and the attacking fighters would hold their fire and allow the bomber crew to either bail out or belly in depending on the extent of injuries to the crew.

I called to the B-17 on his channel to reassure him. "Pick up your gear Big Friend, we're Little Friends." These were the usual nicknames for the fighters and bombers.

"Oh, thank God! My co-pilot is wounded, my navigator is dead, one engine is out, another almost out, my instruments are all shot away and I'm not even sure which way I'm headed". My heart went out to him. I called back and said, "Be sure and tell your gunners not to shoot at us and I'll pass right over your head in the right direction so you can see the heading that will get you home. Be sure that your gunners know we're friendly." My caution was not unwarranted, I had been mistakenly shot at on two previous occasions by B-17 gunners. I held no malice, their bed was one of nails. We got them on course and escorted them back to friendly territory. Since we were a bit short of gas, we then headed for home and landed nearly an hour behind the rest of the gang.

On another occasion, I was leading the squadron when the group headed out on an unusual mission. We were to precede the bombers to the target and theoretically clear the air of enemy aircraft before the bombers got there. Fine, except that since the enemy rarely knew what the target was until the bombers got there, I failed to see the logic.

Anyway the mission called for us to take the usual zig–zag course to the target which was very deep in Germany, Leipzig, I believe. The course in, as usual, was designed primarily to avoid as much flak as possible — if the Germans were also confused about where you were going, so much the better. After reaching the target, the plan called for us to make several criss-cross courses over and around the target. The purpose was to cover the entire area. The last leg was supposed to lead us straight to the target just as the bombers arrived.

It was another solid undercast day. There would be no landmarks or bombers to follow, just a compass heading and time checks to go by. Right at the start of the mission, the group leader aborted with a problem. Soon thereafter the second in command also aborted. It was left up to us three remaining squadron leaders to decide which of us would take over the lead of the group. A two to one vote laid the responsibility on my shoulders. Here I was a wet-eared 23 year old charged with leading a half a hundred P-51s on a complicated combat mission. My gosh, it was only yesterday that I was saving nickels to fly a J-2 Cub. I figured that I was just about to give an exhibition of just how unreliable dead reckoning could be. Fortunately, since I had moved up to occasionally lead the squadron I had given up merely writing the general heading home on the back of my hand. I now carried a plastic covered map with full details, including course headings and time checks written on the plastic cover with crayola. The crayola could be easily erased for the next mission or for security reasons. I was uneasy, but there was no way but to forge ahead with the group following my lead. Not one E/A was to be seen over the cloud covered target area. Eventually the last leg was being flown and what do you know — the bombers appeared right on the nose just like they were supposed to. I had lucked out one more time.

The war was grim and not much to laugh about, but one mission did get me to smile. I was leading a flight of brand new replacements on one of the few ground attack missions we had, supporting our advancing troops. We were loaded with two 500 pound bombs each and the mission called for splitting the group up so individual flights could roam over France and hit targets of opportunity, primarily rolling stock. I was quite proud of my ability to hit targets by skip bombing and eager to show off my skill to the newcomers. I went on my way with my flight of greenhorns and finally found a small village with a fairly large railroad yard and plenty of rolling stock. I brought the flight in from roughly the same direction as the main trunk line. It came in from my right, curved in front of me and entered the yards on the same course as we were on.

Ground attacks were always extremely hazardous, our worst losses and damage came from this sort of mission. Consequently, I opened up as early as possible with the six fifties and sprayed everything and anything that looked like it might shoot back. As I got closer, I concentrated on engines and box cars. It was going so well that I figured I'd let 'em have it with the cannon too, and pressed the button. The instant I hit the button, I realized the error habit can cause. All of my ground

Top left: *Sweet Helen II, the P-51 I flew for most of my tour.* ***Top right:*** *One more shot of "Old Reliable", Stearman N53030.* ***Below left:*** *Home built cropduster built by Herb "JR" Anderson, right, later to become chief of Pitts-Christen Eagle Ops.* ***Below right:*** *The utility A/C, AU-18, test flown by author in 1949.* ***Bottom left:*** *Scott Kemper's Culver Dart, as refered to in the text.* ***Bottom right:*** *Our Luscombe 8-A photographed in 1953.*

***Top left:** Frank Smith's "Mini-Plane" prototype flown by author in 1956. **Top right:** Our Piper Vagabond **Below left:** George Pereira's "Osprey I" being test flown by author in 1959. **Below right:** Another Pereira design, built by author. **Bottom left:** An all too typical landing spot for AG pilots. This aircraft, owned by Frank Michaud, is representative of the larger turbine powered, sophisticated, expensive A/C that replaced our worn out military surplus hardware. **Bottom right:** This is Ernie Stumpf's VP-1 that I test flew for him.*

Top left: *Our Salvay Stark "Skyhopper". Photo by Scott Kemper.* ***Top right:*** *Carl Mueller's Curtis Pusher replica... yes, I flew that too!* ***Below left:*** *I rebuilt this Stits Flutterbug from a wreck. It was a real short field take off champ!* ***Below right:*** *"Junior" Drollinger in his first Lancair at Carson City Nevada.* ***Bottom left:*** *Helen and friend return from a flight in our Ercoupe.* ***Bottom right:*** *The second test flight of the Mace R-2 "Shark", with less than one half hour of flight time!*

Above: *At a fly-in in 1987, Helen and I posed with aircraft types we had originally soloed on many years before. My first solo was in a Rearwin (NC18073) on April 4, 1938 and Helen first flew an Interstate (NC37389) on June 17, 1945.* ***Below left:*** *You can see why there was some skepticism when I first suggested modifying my BD-5 to a tail dragging tractor-engined configuration. Here, at Oshkosh in 1981, Scorchy sits in front of a basically stock BD.* ***Below right:*** *We never really gave it a name - usually called it "Plain Plane"! It's about the same size and performance as a Culver Cadet.* ***Bottom left:*** *The first test flight of our home-built ultralight.* ***Bottom right:*** *A flying ¼ scale model I built of my P-51, "Sweet Helen II".*

gunnery practice was in P-39s. The P-51 didn't have a cannon! I had just prematurely released the bombs! I watched the left one roll and blow up harmlessly in a farmer's field. All I could do was hope that it was brussel sprouts. Fortunately for me, the right one neatly rolled right up on the main trunk line and blew out a very satisfying sized chunk of it. Just a day or two earlier I had escorted a whole box of bombers on a similar mission and blowing a hole in a trunk line was all they had done — I felt quite relieved.

After we got home, I got together with the greenhorns to give them a little encouragement and ask how it went for them. "Everything went OK, Capt. Mace, but I wonder how much practice we will need before we can place a bomb like you can?" I kicked the dirt with my toe and said: "Oh, you'll get the hang of it." I wasn't about to explain that my "accuracy" was a mistake. I smiled a lot that evening.

Some missions where you expected a milk run could be rife with close calls and some losses while other missions where you expected the worst might prove to be a milk run.

The latter was the case one day when I was told no missions of any kind were scheduled anywhere over Europe except for one lone photo reconnaissance by two P-38s over the highly sensitive Peenemunde area. Our squadron was tapped to furnish the two flights of P-51s required to fly escort. It seemed like a suicidal mission to me. Ten fighters would be penetrating deep into Germany over a very sensitive area on a day when there would be absolutely no other activity to divert the defensive forces — surely the defenders would rise up in mass and decimate the intruders. As Operations Officer, it was up to me to decide who would go on the mission. I didn't have the heart to designate the lead to anyone else so I lead the mission myself. We took off and left Leiston for what I feared would surely be the last time. Every mile that went by within reach of the enemy seemed like it would surely bring the end. To my everlasting surprise, not one enemy plane made an appearance and except for the usual flak we completed the mission without a problem.

Being a true aviation enthusiast, I always enjoyed watching flying activity of any kind. So it was that the German night air raids and buzz bombs were a spectator sport to me — no running for shelter. I must admit that the first face to face encounter with an oncoming buzz bomb was a confidence shaker. They were extremely fast, around 400 MPH, and they flew at altitudes of about 400′ to 500′ with long tongues of exhaust flame and gobs of noise! That noise really got my attention as I could hear them coming for at least five minutes before they arrived. That's a long ways away. The coast was lined with British anti-aircraft guns and we were often treated to the spectacle of buzz bombs being destroyed. Those British gunners were good — but not too discriminatory. I watched while they destroyed one buzz bomb in a low fog one morning. About five minutes later I heard an aborting B-17 returning. Then I heard the guns again and rushed outside in time to see a big flash in the fog and the B-17 fall out of it in flames. I thought, "Well, there goes ten Americans." But soon, one by one, parachutes began coming out of the fog. I was told all ten crewmen had survived.

Years later I became good friends with Bert Smeeden, an Englishman selling cars at a British sports car dealership in Sacramento. We were both charter members of the Sacramento Sports Car Club. The subject of the war came up one day and I somewhat mischievously recounted the story. His reaction and reply was a real surprise. "My God, I was the gun Captain on the crew that shot that B-17 down!" It certainly is a small world. I couldn't afford to be too vindictive over the incident. Members of our group had inadvertently shot down two British Mosquitoes — they looked very much like ME-410s to some pilots.

After being forced to fly many hours on instruments, I felt fairly comfortable with my ability to handle any reasonably miserable weather. Underline "reasonably". Not included were nerve jangling times trying to find home base in zero zero weather! At least they didn't seem much better than zero zero at the time and I never felt comfortable with that kind of flying. Once we awoke to weather so dastardly that even the usually cold-hearted mission planners were stymied! They sent a request to our squadron asking for a volunteer to fly up through the horrible mess to see just how high and bad the weather was. No one else raised their hand so, after as long a pause as I thought could be gracefully accepted, I raised mine. I don't know why. Maybe it was because I felt more qualified for the job than the others, or maybe I just hated the idea that someone would think the 362nd was not up to a challenge. Or maybe I was just dumb.

At any rate I soon found myself taxiing out to the end of the runway through muck. When I reached the end of the runway I could see a black mass of rain bearing down on me that was even worse than what I had been taxiing through, it was a gully washer worse than zero zero. In spite of the conditions they cleared me for take off. I quickly countered that I was going to hold for a couple of minutes to see if the mess was going to pass or be a permanent feature for the day. Fortunately, while I sat there agonizing over the stupidity of my volunteering, I was thankfully interrupted by a recall. The big wheels had sensibly decided to scrub all operations for the day and my relief knew no bounds!

My last encounter with enemy aircraft came while escorting bombers to a target in France. The invasion was moving along well and this mission was to help smooth the advance. We were at our usual escorting altitude of 25,000′ to 28,000′ when I spotted a lone speck buzzing the deck heading towards Germany. There wasn't anything going on at our altitude so I reported the bogey to the C.O. and asked for permission to take my flight down to investigate. He gave the O.K. immediately and I started down at a little over 400 MPH. On the way down I was vaguely aware that someone else was reporting a similar find, but I was occupied and didn't pay much attention.

At a lower altitude I was able to make a positive I.D., the bogey was definitely a bandit, an ME-109. He was really hugging the ground, occasionally even below ground level, as he lowered over the creek that wandered in our general direction. I maneuvered into position behind him and was just about to squeeze the trigger when a hale of tracers went by my canopy. I was furious! I knew there was nothing behind me but P-51s and it had to be some over-excited nut trying to steal the kill.

My first priority was to get out of the line of fire, which I did. Not only was I mad but my competitive spirit had been thoroughly aroused. I was determined to get that 109 if at all possible. The greenhorn doing

Another day, another mission. Here we load up a jeep for the trip to the flight line.

the shooting was spraying lead all over the sky and the only way I could point my guns in a hurry was in a side slip. That was no good for accuracy so I wasn't hitting anything either.

Apparently all of this hostility was too much of a distraction for my German. He hit a tree that knocked his right wing off and flipped him over. The 109 stabilized in this inverted attitude just as I slid into pretty close formation with him. I watched as he settled, hit the ground, and his 109 splattered like a broken egg. Ordinarily the pilot's body would have been too hard to spot among all of the scattering rubble, but his parachute had spilled and pointed to him like a big white arrow. Hitting the ground inverted had decapitated him.

I was completely hardened to such sights by now, so all I could think of was finding out who the idiot behind me was. I whipped around in a tight turn to the left and passed over a farmer in a field that was standing no more than a couple of hundred yards from all of the flying bullets and crashing airplane. He was casually resting his hands and chin on the end of a shovel with a total lack of excitement. As mad as I was I had to smile. I guess I wasn't the only one hardened to the happenings of war!

The culprit turned out be the greenhorn who had reported spotting a bogey right after me. It must have been pretty tough for him to see since I had just given full details on its position and altitude! I was angry, not just over the fact that he was trying to steal the kill but that he had to fire right by me to do it. When we got back, I really read the riot act to him. I rarely threw my rank around, practically never really, but there was something about those bullets flying by my head! Being Operations Officer gave me a fair amount of authority so I grounded the perpetrator for a week. That was kind of dumb since all it did was give him a vacation, but it was better for me than biting nails. With all of the confusion, I even failed to make an official claim for the kill. It would have been a perfectly legitimate claim so I now claim it anyway.

The 12th of October mission to Bremen was memorable as it involved an episode including then Lt. Chuck Yeager (later to retire as Brigadier General — first pilot to break the speed of sound) and to some extent myself. The following incidents are exactly as they happened from my standpoint. They are very vivid in my memory. We were leading our respective squadrons and I believe Chuck might have been group leader as well. At any rate we rendezvoused with the 1st Combat Wing Bombers of 3rd Division near Zwolle, not too far from the Zuider Zee. Except for the usual flak, the escorting was going uneventfully. There was an abundance of fighter escort with squadrons of P-51's wandering and weaving around the bombers. I kept my squadron in the near vicinity of the lead box of bombers which was our assigned position. About this time I heard Yeager asking who the fighters were at nine o'clock high to the lead box of bombers. That was my position at the moment so I replied:

"Dollar Leader here, that's our position." "Do you see that group of fighters three o'clock high?" "Yes." Yeager came back with: "I think they're 109s" "I don't think so, I've been keeping close watch on who's who around, and although they are too far over for me to identify from here, I'm certain they are P-51s — what's your position?" Again Yeager: "I'm six o'clock low to that group of fighters, I'm going in closer." Sure enough there was another group of fighters six o'clock low to the fighters in question. "Dollar Leader, they sure look like 109s" I hadn't seen any particular change in positions and repeated my earlier observation. "I doubt it, I'm positive everything around here are P-51s." Then Yeager's final call: "They are 109s! I"m attacking, come over and help!" I still couldn't detect anything to get excited about, but I ordered my squadron to drop tanks and swung over the tops of the bombers. By then I could positively identify all the other flights as friendly and saw nothing happening. The RT (radio) was suddenly full of the typical sounds of fighter pilots in mortal combat but it certainly wasn't around US. Yeager would be credited with five kills in that battle.

I tried to get a better picture of what went on that day by picking the brains of Intelligence. They reported that in forming up, the bombers had inadvertently left a big gap between the third and fourth box. Chuck apparently had taken his squadron quite a ways to the side to investigate something and Intelligence surmised that when he rejoined the bombers, it could have been with the fourth box that looked like the lead box. Since there had been a report of E/A activity in that vicinity, this seemed logical to me, although it is just speculation. I certainly hadn't lost sight of the lead box since the group had formed around them and there was no E/A activity in our area of any kind. Wherever he was, it was certainly a coincidence that groups of fighters would be in the same relative position on two different boxes of bombers. If you talk to Chuck today, you will likely get an entirely different perspective. We certainly were looking through different windows.

Big guns of the 357th included, left to right, Richard Peterson (364th) 16 victories, Leonard "Kit" Carson (362nd) 19 victories, John England (362nd) 18 victories and Clarence "Bud" Anderson (363rd) 16! John England was killed in 1954 in a F-86 crash.

My combat tour was finally coming to an end. After a year of Europe, bullets, and bombs, I was really looking forward to leaving Sweet Helen II and going home to the real thing. The 357th Fighter Group — those "Dirty Yoxford Boys" — had acquitted itself well during its wartime service. In 15 months of combat, 578 E/A were destroyed in the air and 110 on the ground. The 357th owned many records. No other fighter group would destroy as many E/A in a comparable period of combat. Only two other groups would exceed our total number at all, and they were in combat for a longer period of time. No other group would destroy as many aircraft in a single mission — 56, which is a figure not likely to ever be equaled.

The original 27 charter members of my own 362nd Fighter Squadron contributed heavily to the success of the group. Of the 27, eight became aces and four of those eight became double or triple aces. We accounted for at least 128 of the group's 578 E/A destroyed in the air. The price paid was heavy, however. Seven of our number were killed in action by weather, flak, enemy aircraft — or friendly fire. An additional six went down from similar causes but would survive the war. As we left Hayward it would have been hard to imagine that nearly half of our number would go down in combat and that one out of every four of the twenty-seven would not survive the war. But, for every one of those that did die in combat, eighteen or more enemy planes fell under the guns of the Hayward gang. Fittingly, over twenty-one enemy aircraft were shot down by those seven gallant men before they died.

The 362nd Fighter Squadron Pilot's Nissen hut showing squadron emblem above Eighth Air Force insignia on fireplace and a partial victory tally on wall.

In no way do I wish to ignore, slight or diminish the accomplishments of those pilots that joined our outfit after we arrived in England. Certainly they figured heavily in the success of the Group. But the hectic, hazardous, trying times of combat did little to equal the opportunity for forming the bonds and kinship that the many months of training in the United States had done, at least for me. Furthermore, it was only natural for the old gang to bunk together in the same Nissen huts at Leiston and that gave less opportunity to get to know the new pilots. Indeed, since a significant number would not survive the war, it really seemed better not to. As a result, of all the replacement pilots we had in our squadron, the only one that I really got to know as a friend was Chuck Weaver and that was probably because he was often my wingman when I became a flight leader. The rest just seem to fade from my memory but their successes are forever etched in the history of the Group. Many, like Chuck, went on to become aces, flight leaders, or both.

For my part, I had flown 59 of the group's 318 combat missions and in nine of the group's 15 months of combat service. Twenty-eight of my 59 missions were over 4.5 hours, 18 of which were over five hours, two over six hours, and two over seven hours — and one of those was nearly eight hours long. Although I had downed only three E/A, my batting average was 100% while never dodging a fight. For my efforts, my fears, my hardships, and my numb behind, the Air Force awarded me the Distinguished Flying Cross and seven Air Medals.

In evaluating the P-51's fighting ability, it can only deserve an A+. It had no vices, and was easy to fly, and its overall performance was unequaled. Perhaps the Spitfire could outturn it, the Typhoon out-run it, and the Thunderbolt out-dive it, maybe, but no other fighter plane could put it all together like the P-51. More importantly, no one will ever know how many bomber crew lives were saved because of their "Little Friend," the P-51 Mustang.

LIFE AT HEADQUARTERS

Home was going to look awfully good. Ah, but you just can't rely on the service. I was ordered to report to 3rd bomb Division Headquarters. They needed a Fighter Controller. "Why me? I'm just an ordinary captain who just finished nearly a year of combat in fighters!" "That's exactly the qualifications they require and at this time you are the only one in England who qualifies." That was hard to believe.

There was no choice. I reported to Elveden Hall, 3rd Bomb Division Headquarters, General E. E. Partridge, presiding. it was not all bad. Except for not getting to go home, it was hog heaven, actually. It was staffed by ex-combat bomber crew members who had done most, if not all, of their tours before we came over with the P-51s. And if any combat bombing is bad, unescorted combat was pure hell. Bomber crew members looked upon us fighter pilots as saviors. I was not just another employee, I was an honored guest! It opened a lot of doors for me and one of the first things I did was check on that business of my being the only fighter pilot in England meeting those meager Fighter Controller requirements — sure enough, at that time I was. It was unbelievable. However, the living quarters were sumptuous, the food excellent, and the duty meager with gracious comrades.

Shortly after being assigned to 3rd Bomb Division Headquarters some higher up must have decided it would enhance my value to the operation if my horizons were broadened. As a result I was directed to spend a week at Pinetree, the super secret, highly guarded underground headquarters near London. Unfortunately answers to my questions of what I would be doing there ranged from being extremely vague to nonexistent. So, after gathering a few essentials, I found my way down to my destination still pretty much in the dark about my coming duties. My papers were processed and an ID pass was pinned to my chest that let me into the hallowed recesses of this underground building. I hadn't seen anything so heavily guarded since those few empty blocks at Camp Hahn! There were guards on every floor and in front of the door to every room. No room could be entered until the guard knocked on the door and had some one from inside come to the door and O.K. your entrance. The pass on my chest must have been a powerful one because I was never denied entrance no matter where I went, or what room I picked to enter.

A photo of my return to Leisten in a P-47 after joining the Third Bomb Division as Fighter Controller.

The only trouble was that no one there seemed to have any more idea of why I was there than I did! There was no Liaison officer, guide, or friendly G.I. to show me around, and certainly no Information Desk! I only knew that someone said I should go there and I now had to kill a few days doing nothing while appearing to be doing something. I wandered from floor to floor and into room after room where my presence was totally ignored, or as in one case or two, where my presence caused a slight chill in the atmosphere by the Officers gathered in conference. In the latter case I simply slunk back out and tried another room. In no case did I ever come away with anything that would have compromised the war effort had I run to the enemy with it. It was a very boring week and I was very glad to get back to Elveden Hall. When I did get back I was never questioned about the venture, it was as if I had never been gone.

The Fighter controller's job required waiting for 8th Air Force to send down the bomber missions over the teletype each evening. He then assigned the appropriate fighter cover and put the orders out on the teletype to the fighter groups involved — and that was it. The whole procedure took about twenty minutes. The teletypes usually arrived anywhere from early evening to midnight. The days were ours though, and to make the assignment even better there were two of us. A Major was already there so we were only on duty every other night. We could even trade off if we wanted to — three days on, three days off, or what ever — it was strictly between us.

Just so we wouldn't get bored — actually so no one would miss their flight pay — we (Headquarters) had our own field and a fair assortment of planes at our beck and call. Two to four each of L-4s, Cessna Bobcats, AT-6s,

P-47s, P-51s, and B-17s, were all gassed, ready to take whenever or wherever we wanted to! It may seem like a terrible waste, but not to me. Not at all, those ex-bomber combat crews could never be adequately repaid for what they endured.

The head navigator in the Headquarters had stood down one day, for instance, and the squadron suffered a 100% loss. Gathering spares and replacement planes, the squadron was sent out on the next mission and all were lost but the navigator's aircraft. Very poor odds indeed.

On another occasion one of the pilots suffered severe hits on his plane from a head-on attack. A 20 mm shell hit his copilot in the neck, severing his head which bounced off the bulkhead and landed back in his lap. The pilot had to suffer this spectacle for the rest of the mission wondering if his turn was next.

Shortly after my arrival at 3rd Bomb division, some of the bomber crew people decided that witnessing a bombing mission from the perspective of a fighter pilot would be very interesting. Since they had an experienced fighter pilot, me, and a two place P-51 available, they began to badger me to take them on a mission. I had no desire to participate in the scheme for numerous reasons. For one, I had done my tour and felt enough was enough. For another, the two place P-51 was unarmed. I kept being badgered, however, until the point was reached where I either had to take them on a mission, or admit to being a coward. I finally just bit my lip and agreed to take them.

Since the head navigator was the most eager, he was the first in line. I piled him in the back of one of our AT-6s and flew to the field where the P-51 was located. One of the things that a veteran fighter pilot never loses is the need to keep looking around. So it was that upon nearing the field I spotted a P-51 bouncing us from five o'clock high. I immediately accepted the challenge and, by putting the AT's superior maneuverability to work, I bested the P-51 at the expense of some vomiting in the rear cockpit. I landed immediately, feeling relieved that undoubtedly my passenger would want to call the whole thing off.

But, no, after a few minutes recovery, he insisted on going ahead with the mission. We climbed aboard the P-51 and headed out to rendezvous with the bombers. As we neared the bombers I spotted a lone fighter quite close to them. It was too early for any fighters to be joining them so I immediately closed in to make sure it was not hostile. This necessitated much wing rocking to assure the bombers we were friendly — which precipitated more vomiting behind me! The fighter turned out to be a P-51 and when I slid in close I was able to identify the pilot as General Partridge. Sneaking out alone to observe the bombers was a bad habit of his.

A side light to my job was to keep an eye out for this activity of his and assign fighter escort for the General whenever we caught him at it. I called up some escort for him over the RT and checked on my passenger. He had used up the last of my maps as a place to store his recycled breakfast and now he finally was ready to go home! As I headed back I thought I should at least give him a little bit more to talk about, so I put the nose down and poured on enough coal to nudge the airspeed up to 400 MPH indicated, something he had never seen in a bomber. I leaned to one side and pointed to the airspeed indicator. He leaned forward to look — and puked in his hat. No one ever asked to take them on a mission again!

One of the airplanes I was extremely eager to fly was a P-47 Thunderbolt. There was no one around to give me any pointers but I couldn't wait and decided to take one up anyway. The only trouble was that I couldn't be real sure about how to start it. Luckily a knowledgeable mechanic strolled by and I got him to show me.

If I thought a P-51 was blind in three point attitude, I hadn't seen anything yet. With that big fat nose, and the cowl flaps out, half the horizon disappeared. The P-47 proved to be a smooth, solid old tub, however. For a comparison, I would say the P-51 was a hot Model "A" and the P-47 a Cadillac. It was rugged. I really enjoyed it and flew it many times.

I was also checked out and qualified as first pilot on the B-17. Except for handling the controls of a B-24 for about five minutes in Idaho, I had no multi-engine experience and wasn't about to jump into this sort of thing without a good checkout. The pilot who checked me out was a young captain. When I got to the field the ceiling was down to about 200′ so I said, "No go, huh?" He said, "Oh, it's all right." I soon found myself in the left seat taxiing out to take off without even so much as a familiarization ride. Those combat B-17 pilots were blase! As soon as I got airborne and the gear up, he said, "Go ahead and fly it around a while." He then got up and went to the back to shoot the bull with some G. I. 's who had come along for the ride. That was the last I saw of him for about twenty minutes. There I was, an absolutely green multi-engine pilot, making turns in a completely strange airplane with one wing tip nearly in the fog and the other just missing the trees with some guys in the back who couldn't care less. I'd met my match.

The B-17 was a fine flying airplane, I later flew it on instruments and made simulated three, two, and single engine landings, which were all easily done. I thought its landing characteristics were quite similar to the P-47.

In a moment of complete mental lapse I even volunteered to ride along as an observer on a combat mission. Well, it was pretty much of a milk run. The mission called for a reconnaissance run just barely over enemy territory by a single B-17 manned by a brand new crew. After many hours of flying, and not much going on, I was half asleep behind the navigator when one of the crew screamed over the intercom:

"There's a jet on our tail! A jet on our tail! We're going to get shot down! " Needless to say I became thoroughly awake. There was a moment of confusion until the pilot remembered I was aboard.

"Let Capt. Mace up in the bubble so he can evaluate the situation." I quickly took a look to the rear and saw nothing.

"Where's the jet?"

"Down there, look down there."

"Down there" was about 5,000′ below us and the "Jet" was a V-2 going straight up; it would come close but hardly "on our tail." There was considerable relief when I gave my report. The V-2 was on its way to London and did pass close enough to read the serial numbers however. The V-2 was entirely different than the V-1. Its approach to the target was straight down and supersonic. It was seldom seen, if ever, and never heard until it exploded. If you heard the explosion, then you knew you were not a victim. Only after the explosion could its approach be heard. The noise was not unlike that of a freight train. I've had the "pleasure" of being in the neighborhood when they hit on a couple of occasions.

Eventually it was time to get the B-17 home. Naturally, being late in the day, it wasn't too surprising to find England completely engulfed in zero-zero fog. Fortunately the powers that be had made arrangements to save the day under such circumstances. A very long and wide field at Manston was an emergency field. The runway was completely bordered by pipes and jets, fed by gasoline. When lit, the heat was supposed to raise the ceiling of the fog over the runway to acceptable levels. From our vantage point, there was just a smooth low blanket of white — except right over Manston where there was a neat, raised, rectangular shape of the runway. All we had to do was aim an approach at the end of the raised cloud, and except for a distinct bump from the thermal off the fire at the end of the runway, it worked as planned — with a huge cost in gasoline! it was called FIDO for Fog, Intense, Dispersal Operation.

While at 3rd Bomb Division, I became acquainted with a colonel who apparently was in charge of a new secret bombing procedure called Project Aphrodite. The idea was to take a stripped B-17 and equip it with radio control in addition to the pilot's controls. It was then filled to the brim with high explosives. A pilot was required to take it off and climb to a safe altitude where a radio control pilot in the mother ship could take over and allow the takeoff pilot to bail out over England. It was also equipped with a television system which permitted the radio control pilot to fly it below a cloud layer and still see a target to dive in on.

The Colonel asked me if I wanted to volunteer to take the next one off and bail out. The whole thing sounded interesting so I quickly agreed to. I then asked. "How many missions have you had so far?" "Four," he replied. Jokingly I said: "And no fatalities, I presume." "Uh, well yes there has." Now I was getting uneasy. "How many?" "Uh, well, uh three pilots." was his quiet reply. "Three pilots!! Three pilots out of four missions?!", "Well, it's not as bad as it seems, on the first mission we had a pilot and copilot aboard, so that one accident accounted for two pilots." Even a simpleton should easily see that it still added up to a 50% pilot loss — and I'm not a simpleton. "Sorry, Colonel, I just de-volunteered." I won't go into the somewhat embarrassing, and probably still secret results of this effort — suffice it to say that radio controlled flying bombs did not figure heavily in deciding the outcome of the war. For many years afterwards I thought it was the same operation that killed President Kennedy's brother, Joe. But I now believe he was killed in a parallel operation involving B-24s and the Navy. Evidently theirs didn't work too well either.

Having all of these wonderful airplanes to fly was certainly a potion to kill some of the pain of not going home. In spite of having all the heavy stuff to fly, I still enjoyed flying the good old Piper Cubs. One of the other pilots and I were up dogfighting in the Cubs one day until I noticed our gyrations had taken us down to about 100′ over someone's bomb dump. Discretion seemed to advise scooting over to safer ground!

It also led to three days of more than passing interest. One of the Colonels stopped me in the hall one day and asked if I would mind going over to France to pick up one of the P-47s that he had to leave over there to get fixed. It seems he had taxied into a bomb hole and bent the prop. It sounded like a fun trip, so with only passing curiosity as to why he wasn't eager to go back himself, I quickly agreed. I rounded up another pilot to bring back the AT-6 I was going to use for the trip over. The P-47 was located at a small town well north of Paris and very close to the front lines. Crossing the Channel on the way over I casually mentioned to the other pilot how interesting it might be to take a look at Paris. The enthusiasm from the rear cockpit dictated an immediate course change to Paris. The weather was beautiful so to get a close look at the Eiffel Tower, I executed a racing type 360° turn around the tower. It was still possible to get away with such things at that time.

After we landed, we found transportation into Paris where we planned to see what the highly publicized night life in this big city was all about. It was already getting dark when we began touring the night club area of the city. We finally came across one of the fancier looking night spots and decided to give it a try. We no sooner were seated at a table than two somewhat brassy looking broads came sailing over and invited themselves to sit down and join us. I was — and am — pretty naive about night club night life, but I was hip enough to spot shills when I see them, so their effort to extract any money out of me was a waste of time. Particularly since I wasn't carrying that much anyway.

I did have enough that buying a couple glasses of apple juice wasn't going to break me so it wasn't long until I had to relieve myself. The restrooms were labeled in French but that was all right and I breezed right through the Monsieur door only to end up face to face with a woman! I beat a very hasty retreat to reread the door labels. I just had to be right but decided to wait and see what door the next person used, I really had to go. The next prospective user was a Frenchman who also breezed right through the Monsieur door and didn't rush right out again. Since the French culture can be pretty strange that still didn't convince me. My up bringing had made me extremely shy and even the military's hard nosed approach hadn't changed my attitude that much — I decided to wait for more confirmation. Soon a G.I. sailed through the same door and that was finally enough to convince me. It turned out that the woman was the rest room attendant! Talk about culture shock! I don't know how long it would take me to get used to a strange woman brushing the lint off my back while I'm taking a leak!

Back at the table, my peaceful resumption of people watching was soon interrupted by two M.P.'s tapping us on the shoulder and informing us that we were under arrest! We soon found ourselves on the way to the pokey wondering what was going on. In short order it was all laid out for us by a hard-nosed iron backed O.D. It seems that to offset a near runaway AWOL problem at Paris, the city would periodically be declared off limits to any military personnel except the MP's. They would then comb the city and arrest everyone in uniform. That was the first we had heard about such goings on. Since we were without any written orders, it took considerable effort to convince the O.D. that we were on an official mission VOCG (vocal orders of the Commanding General) — well, a simple request by a Lt. Colonel is pretty close! The O.D. did relent and released us providing we got out of town in 30 minutes. Well, we had seen enough anyway.

The sad end of a gallant warrior... Kit Carson's beautiful "Nooky Booky IV" being broken up in Germany after the war.

Back in England I had heard rumors that a tidy profit could be made by exchanging strong English money for weak French money outside official exchange places. It was highly illegal, of course, and the high profits quoted didn't seem likely to be true to me. I've always been an extremely honest person — but also very curious. So when a seedy looking character approached us on the way back to the airport waiving Francs at us, it didn't take a genius to figure out that we had found one of the black marketeers and I decided to satisfy my curiosity. I traded a Pound note for a bundle of Francs and sure enough the profit turned out to be pretty tidy indeed. It was tempting but I let that be my first, and last, excursion into crime.

We took off next morning and as I was nearing the town of Nancy where the P-47 was, I spotted a B-24 bellied in on a field. It looked so fresh that I thought the crew might still be in it so I decided to go down and take a look. As I got closer I caught the glimpse of an ME-109 on the deck headed right at us. There was no time to wonder where or why, I just immediately executed some violent evasive action. After enough time passed that he should have gone by, I took a hard look around but found no 109. I looked back to where I saw him in the first place and he was still there — on his belly. It was kind of a shame that the German pilot wasn't still around to see my beautiful display of evasive action! With the old heart pumping a little faster we continued on to where the P-47 was located.

When the field came into view, I had the first inkling of why the Colonel wasn't too anxious to come back for it himself. The field was grass, and very short. Upon landing, it also proved to be quite soft from all the rain. The whole deal was looking less and less attractive. To top it off, the prop on the P-47 had been replaced by the only prop they could find. It was one of the older needle blades instead of the more efficient paddle blades. Well, I had said I'd bring it back so I sent my partner off to England in the AT-6, preflighted the unarmed P-47 and climbed in and taxied out. While turning around at the end of the runway, the right wheel sank in the mud and I was stuck. It evidently was no surprise because several G.I.'s quickly appeared and lifted on the wing enough to get the wheel out — only to have the other one sink. After a couple more times of this, I suggested that if they could stand the gale from the prop wash I would just blast off the instant they got the wheel out the next time. Heads quickly nodded affirmatively — I think they were anxious to get rid of me. The plan worked, and at last I was rolling down the "runway". With takeoff flaps down, it looked like we were going to get off the ground without too much trouble. But just as the P-47 got light, one wing started up and a quick glance at the flaps revealed the problem.

This was an older P-47 with flaps that occasionally didn't stay down evenly. To counter this, Republic had installed an equalizer valve that was inconveniently located somewhere under the seat. There just wasn't time to find the unfamiliar valve so I dumped the flaps instead. This meant that I would now be using up what little cushion I had in runway length. The old 47 and I were going to be pretty close, but we made it... but only because the grass field sloped down at the far end! Heading for home, I realized Brussels was not far off course, and, surely I deserved a little more of a breather after what I'd been through. So it was off to Brussels. I landed and taxied by a very familiar looking AT-6.

A little later we got together and, after determining that it was not off limits, decided to try the night club routine in Brussels. We toured the night club area and came across a small unusual looking place. It featured a Halloween like atmosphere, even the entrance was very unusual in that they made it look like you were entering a cave. You had to stoop to get through it and the interior was also suitably decorated. It also drew a wild assortment of clientele that made for good people-watching.

Once more I resorted to my apple juice and once more reached the point where I just had to get rid of some. Unlike the Paris club the location of the rest rooms in this darkened little night spot were nowhere to be seen. There was no alternative but to ask the matronly waitress where it was. This presented quite a problem because she spoke no English and my meager French was just not going to get the job done. As a result, my dire need demanded that I get the message across by using sign language alone! Somehow I managed to succeed without embarrassing myself too much and she laughingly motioned for me to follow her. She led me to the restroom — and to the urinal — and just stood there!

Just why she felt the need to stand there was beyond me (I can only assume that she wanted to show me back to the table). After waiting long enough to assure myself that she was definitely not leaving, I went ahead. The toilets were in a larger "L" shaped part of the room and the urinals were lined up against the wall of a narrow hall section. Peeing in front of this matronly gal was bad enough, but just after I finally got going two young gals came in and walked down the narrow hall behind me side by side. One bumped me with a shoulder that nearly put me into the urinal. They just walked on talking a mile a minute! Needless to say it was shocking to find out that that was the only entrance to their toilets! Europeans are certainly blase about nature's needs! But, I suppose their attitude is better than the uptight American approach.

Eventually curfew time arrived and the club was closing up. With no place lined up for my partner and I to bed down for the night we stood there looking a little lost when the matronly gal came over and, with difficulty, managed to get it across that she and the bar tender owned the place and would like to have us join them upstairs. That seemed mighty friendly of them so we accepted the invitation (how can you turn down a lady that has watched you pee?) and we all trooped upstairs and sat down. That is the lady, my partner, and I sat down; the old man disappeared in another room. After serving us a couple of apple juices she held out her hand for money.

I didn't have to be hit with a Mack Truck. It was obvious that friendship had nothing to do with inviting us up, we were there to be fleeced. I made a "lets leave" motion to my partner which the gal also quickly picked up on and, in a desperation move to keep from losing her lambs, quickly came over and plopped her oversized butt on my knees and stirred a hard elbow in my crotch. Now, I can't speak for all 23 year old guys, but I personally don't consider this as a turn on.

Next morning a G.I. was helping me roll the bird out and when he found out where I'd come from and what time I had left the day before. "Boy, you were sure lucky." he commented. "Yes, the field was pretty short," I replied, surprised that he would know that. "Very lucky indeed." After a couple of more "Boy, you were sure lucky's" I wondered if he knew something I didn't, so I asked him. He did. "My gosh, haven't you heard? The Germans counter attacked yesterday and that town and field fell 45 minutes after you left." So began the Battle of the Bulge... and the day wasn't over yet!

I fired up and taxied out to the end of the active runway. As I was running up for the mag check, my ears picked up a familiar rumble — buzz bomb! I looked over my shoulder and at no more than 400' high came a buzz bomb right over my head. To make it even more interesting, the local big anti-aircraft guns opened up and great big blobs of black flak were bursting right over me. Luckily they didn't hit the buzz bomb or I might have been in quite a pickle. As it was they managed to put a small puncture in my left tire, but due to a very hasty departure, it wouldn't be apparent until I landed back in England. The tire was flat by then but fortunately I was able to easily keep the old bird straight on landing. Since taxiing was out of the question, I parked it just off the runway to end a very interesting three days.

The war was winding down, and memories of the warm comfort of Helen were really getting to me by this time. I decided to lean on the esteem for fighter pilots that my bomber crew superiors had and badgered my way home to the States. Even the offer of an immediate promotion to Major — providing I agreed to stay — did not sway my decision to leave. So it was that a ride on the hard seat of a C-54 brought the irreplaceable United States into view once again.

Helen and I had our most memorable reunion and long delayed honeymoon at a Los Angeles Beach Hotel, thoughtfully provided by Uncle Sam. It couldn't have been nicer — right on the beaches that we virtually grew up on, and under the departing planes from good old Mines Field. A most satisfying R&R it was. All too soon we were off to what would turn out to be our very last assignment in the U.S. Air Force, Bryan Army Instrument Instructor's School. What was this? Some monster joke?! Punishment for some unknown villainous act? It couldn't be! Visions of the incompetent instructors screaming in my ears during the drudgery, sweat, and pain of those few hours of instrument training as a cadet filled my mind.

However, I was in for a pleasant surprise. The purpose of the school was to turn out instrument instructors, not fodder for the war. It was extremely well staffed with very professional instructors and I was pleased to have gone through it. Also I had applied myself to the task in view of the penalty for incompetence, a lesson hard won over Europe. I was even asked to stay on as an instructor, and was giving it serious consideration when peace dictated the dismantling of the greatest Air Force the world had ever seen — or might ever see. The civilian world was about to be flooded with hero pilots and also rans — and hams in both categories.

Hamming it up in an airplane is an irresistible urge to some of us. It certainly should be tempered with caution and common sense. Who could forget one of our young replacement hotshots who celebrated the finish of a very successful combat tour with a low altitude victory roll — right into the ground. Or the buzz job of a B-24 that ended with a steep pull-up and stall that threw the only survivor out of the waist window before the crash. Or the P-38 pilot who buzzed our field so low that he damaged one prop — then in his excitement feathered the good one. Or the very famous WWII Ace (I'll leave nameless to spare him further embarrassment) who made a tour ending celebration buzz job even lower than the P-38 — and wound up with what surely is the world's fastest belly landing with survivor.

While we were at Bryan the opportunity was again available for Helen to continue learning to fly. A small FBO in the area had a couple of Interstate Cadets that were used for instructing purposes. These were not all that different from what I had soloed in before the war and were ideal for Helen. The Interstate was the first of a wide variety of airplanes that Helen would eventually fly. That variety would run close to 40 different types of standard and experimental airplanes, from Cubs to a big five place Waco cabin.

Helen's Viewpoint

After my brief start at learning to fly in Pocatello, I was really eager for the time to come when I could continue with flight instruction. While Harvey was overseas, I took the aeronautical night classes at UCLA to prepare for the written part of a Private License, a goal I was hoping to reach.

Being at Bryan gave me the wonderful opportunity to start taking lessons again. It was good getting back in the blue with a really great instructor. When I took off for my first solo flight, I seemed to hear his voice guiding me through to the landing. What a gold star day that was! I felt very fortunate, and Harvey was thrilled over my soloing.

CROP DUSTING AND OTHER CLOSE CALLS

Of course, without the benefit of the salary of an Air Force Captain with flight pay, some means of putting food on the table would have to be found. I wanted to keep flying and first considered the airlines. At that time co-pilot's pay was very poor, and transfers to the left seat were apparently years away, so I quickly gave that up. Back at Dycer, before the war, my boss Elwood had strongly opposed the idea of crop dusting as a way of making a living. I hadn't forgotten his well intended advice; but in our travels around the south, I had gotten glimpses of it now and then and liked what I saw. I decided to check into it. Besides I just couldn't hack the idea of the nine to five route. I found a couple of crop duster pilots who were talkative (I would find out later that they all are!) and sure enough big paydays could be had — if not too regularly.

We decided to go for it and asked for jobs with a lot of operators in the south, and most of the ones in California and Texas, and got an almost universal reply — some almost belligerent, "You Air Force pilots don't know how to fly, it takes extra skills to fly one of these things." "Get lost" was the unsaid finish to their reply. I tend to be persistent, though, and finally in Carrizo Springs, Texas, we found an operator whose alcoholic pilot was giving him problems. Since Carrizo Springs wasn't running over with pilots, I was hired to replace him. He had a J-3 crop duster and that suited me just fine, so what do you know, military pilots could fly those things after all.

Of course there were a bundle of questions about crop dusting that needed answering. Not the least of which was "What does a crop of tomatoes look like?" — I'm the city boy that didn't even know what a sod field was! There were plenty of technical questions, too, like when do you open the hopper door? Do you have to lead the edge of the field like dropping a bomb? How large an opening do you set the hopper door for? How wide do you make the passes? How much wind are you permitted to work in? How high — or low — do you fly the passes?

Since this operator did dusting only — no spraying, seeding, or baiting — my initial learning task was made much more simple. The dust needed no leading, just open and close the hopper door at the edge of the field. The boss set the door opening, so any small adjustment of material applied could be made with the width of the passes. The height of passes would generally be from one foot to perhaps five feet, regulated in part by the small amount of wind drift that was permitted. Of course, when there was drift, thought had to be given about what might be in its path down wind. In this area, only one type of dust was being used, a mixture of arsenic and sulphur. You had to be especially careful with bee keepers around. The bees would almost expire at the sound of the word arsenic.

Whenever possible, further assurance of getting even coverage of the field was accomplished by splitting the field in two or more even sections. Running a little short on the first part meant spreading the balance out a little further. A couple of headland, or cross passes, at the end of the job, covered the ends of the field for missed spots because of pull-ups over trees or whatever.

Things went very well, except that the season that had paid the boss so handsomely the year before was falling on its face. He had badly over extended himself financially and was going broke. By trading the big Buick that Helen and I had at the time along with some bucks out of our bank account, we suddenly found ourselves owners of a failing crop dusting business with a J-3 duster. We milked the rest of the season for whatever was left while Helen once again filled the financial gap by teaching school.

Mentioning the Buick reminded me of the time we were returning from Eagle Pass. About 25 miles from Carrizo Springs, Helen and I passed a car along side the road with the hood up and two guys looking a bit down. We backed up and asked if we could help. These were still days that were not fully recovered from the war and traffic was scarce everywhere, much less in this part of Texas. Their car problem was insoluble but they said they would sure appreciate a ride to town. "Heck, get in your car and I'll push you home." I replied. "Twenty-five miles?" "Sure, why not?" Always the optimist. That old Buick did a fine job. The guys might have preferred a little slower speed but they seemed grateful anyway.

Several days later we had a job to do too far from the field to ferry the loads from the airport, so we loaded and flew off the highway near the farm. This was not an uncommon procedure in remote areas. This road, like most roads in those days, was narrow and we had to take down several mail boxes to make room. When it came time to leave, the boss got them all up again except one that he overlooked. It wasn't long before I got word that there was a complaint and the Highway Patrol was "looking for that cropduster who was using the highway for an airport!" Days passed with my being fearful of being apprehended every time I ventured out. But after a while I began to wonder why the Highway Patrol could not find me in this dinky town, until one day I caught a glimpse of the troopers. They were the same two guys Helen and I had pushed the 25 miles home! Talk about removing a thorn from a lion's paw!

There was time enough between jobs for things to get a little boring so occasionally Helen and I would fly out to one of the more remote farms to visit the folks there. Since the Cub duster had only one rather small seat, it meant squeezing Helen in on my lap. It was cramped, but we managed. Landing that way on short narrow farm driveways was a bit of a challenge, however.

Describing this country as a farming area might give folks from the mid-west or California the wrong idea of what the area looked like. It was actually very primitive desert landscape inhabited by coyotes, cougars, wild boar, and rattlesnakes — big rattlesnakes. The farms were just plots of cleared ground here and there, irrigated with well water held in man-made diked up lakes called "tanks." Many of the homes didn't have regular electricity. They relied on wind driven DC generators and large banks of batteries.

It was Indian country before the white man came. One couple we visited took us around to undeveloped parts of their ranch and we were able to see the remains of Indian activity. There were little mounds here and there where the Indians had made arrow heads. There were many pieces of arrow heads and quite a few undamaged ones scattered around. It was all very eye opening to this city boy!

On ferry flights I occasionally spotted little herds of wild boar that I couldn't resist buzzing. Their wild eyed look, wire like hair standing straight up on their backs, and big curling fangs made them look awfully mean. I just had to hope that I wouldn't have a forced landing after antagonizing them. Of course, there were lots of other reasons for not having a forced landing in that country.

This part of Texas was a lonely place so when the season gave its last gasp, we decided to head for California. As we got ready for the trip, I noticed the lack of a carburetor heater on the Cub's engine and briefly wondered if that was going to be a problem going home. I found out very quickly. Not twenty miles down the road, she iced up and quit. The highway was long, flat, and straight so I sat it down on the

Left: Our first Cub Duster, Carrizo Springs, Texas. Right: Helen flying our fourth airplane, Piper J3-65.

road and pushed it off, tail near the fence and nose near the pavement and waited for the engine heat to soak down and Helen to come along with the 1937 Ford we had ended up with.

I couldn't understand it. I'd been flying the Cub for weeks in this same area without a sign of ice, yet the first thought given to it brought trouble. While waiting for Helen to let her know we would have to watch for this on the rest of the way home, one of the locals came by, stopped, got out and walked around surveying the whole thing. Finally, he ambled over and asked, "Did you just land that thing?" "Yes, sure did." "Goll-lee, had no idea you could land them things so short." I looked up and down the approximate two miles of straight highway. "No problem huh? Wow, that's really sumthin!" He got in his car and left. Only then did I realize that he thought I had landed it where she sat!

Cubs can be landed very short — but not that short. Being able to land short is a major requirement of crop duster pilots, however. Job demands will often dictate landing and taking off in places Sunday pilots wouldn't dream of — highways, roads, or truck access paths with crops on one side and a ditch on the other. Very often you're required to scoot under power and phone lines on either end because they are too close to go over. All of this often had to be done with vision impaired from sulphur in the eyes. I've landed many times on farmer's driveways to collect or check on fields. As might be surmised by now, being a crop-dusting pilot was somewhat of an outlaw affair. In my experience it certainly was anyway. Where practical you went by the rules, but if it wasn't practical, and you thought you could get away with it, what the hey!

Helen soon arrived on the scene, and after a short consultation, we were on our way — briefly. It was only a few miles to the next forced landing, again on the highway. There would be seven forced landings before we reached Los Angeles. The closest call would be when I was trying to sneak through a mountain pass under a very low ceiling. While traversing a short valley between two ridges and approaching the far side, I detected the early signs of carb ice. The road didn't look usable because of the terrain but I had passed over a flat field next to it about 150′ to 200′ long. That is long enough for a Cub, but just as I throttled back to land I spotted a barbed wire fence running right across the middle of it — that's when the engine decided to quit. The only other choice was the road. It ran flat across the valley for about 350′ — but there was a bridge too narrow to get through near the middle of it. Well, it was more than the field had to offer, but it would be very marginal. Luckily there was a wide shoulder on the approach end of the bridge where I could make a controlled ground loop to keep from hitting the bridge. A slight tail wind had also aggravated the situation.

Traffic was very sparse in those days but before Helen arrived a young Air Force pilot drove by, slammed on the brakes, got out without saying a word, and after surveying the whole scene asked, "Where in the Hell did you land that thing to get here?"

I was getting increasingly concerned about the Feds since we were getting closer to civilization. When Helen finally arrived, I hurriedly piled into the car. "Let's check the summit for weather and see if it's passable so we can get out of here before the C.A.A. catches us and hangs me to the nearest tree." "Oh, don't rush it! There's not going to be any C.A.A. way out here." We drove on to the ridge anyway and the weather looked passable so we hurried back to the Cub — and to the sight of an individual leaning on the strut and a black panel truck parked alongside with the government insignia of the C.A.A. on the side. Oh No!

As it turned out, he was a very easy going guy and after determining that we didn't need help — "Oh, no we're just waiting for the weather to clear up, we don't want to fly illegally..." he drove quietly away.

That still left me facing a very trying takeoff situation. The best bet would be opposite to the way I had landed since there was a very light head wind in that direction. It would be very close. Backing the tail as far into the bridge as possible, I held the brakes and gave full throttle. After a reassuring nod to Helen, I let go of the brakes. The old Cub came through and lifted off barely in time. After a very slow speed turn to dodge the rising terrain, we were once more on our way.

Airports were few and far between in those days necessitating much improvisation along the way. By using the outdated maps that were the only thing available, I would occasionally arrive at my destination low on fuel and find no sign of anything remotely resembling an airport that the map had shown. Or, on the other hand, a destination would have to be picked where we knew there would be no airport and we'd just have to hope some spot was available to land on. Usually I stayed close to the highway but one stretch of highway arced entirely too far out of the way to follow. The crow's path was over desert with the destination beyond the range of the Cub so I filled a five gallon gas can I carried in the hopper with car gas, the only thing available for most of the trip, and hoped to find somewhere in the desert to land and refuel. If anything dire had happened out in that desert, I don't know whether I would have ever been found or not. Luckily, nothing did.

Since Helen was already a qualified solo pilot, she flew an occasional leg herself and eventually we wound up back in good old Los Angeles.

Having decided to pursue the cropduster's way of life, I figured that it would be extremely advantageous to have an A&E license. The G.I. Bill was handy to make use of so with Helen working once again, I attended Cal Aero Technical Institute full time for the next year and graduated as a licensed A&E Mechanic. It proved to be a good move.

Cal Aero and the Navy Mechanics School, where Helen had found a job, were both located on the Grand Central Air Terminal in San Fernando Valley. It was a long commute twice a day from her mother's place in the south part of Los Angeles. Naturally it was right at the busy commute hours and before there were any freeways. At least we were able to suffer through it together.

Helen with our second airplane, an Aeronca L3-B.

To help the finances we sold the duster but had to take an Aeronca L-3 Defender as part trade. At least the switch gave us each a seat to sit in while we flew together. We kept the L-3 at Compton Airport and while we were there, we found the old Wiley Post that had been at Dycer before the war. It was sitting there, forlornly, with a broken lower right wing from wind damage. The owner was begging someone, anyone, to take it off his hands for $150. Eat your hearts out Antiquers, but remember this was before the Antique restorer's bug had spread. It wasn't an Antique then, just an old Model "A" powered biplane with a broken wing. We couldn't resist and bought it.

We repaired it and flew it regularly for a while. It was a real fun airplane. occasionally we even commuted to Cal Aero with it. The engine was getting tired and when a coke bottle I had set on one of the upper wings fell over and went through the fabric, we decided it was time to move it. We had to practically beg someone to buy it for $350. We still had the L-3 anyway.

Finding a cropdusting job after graduation took a little time, but the operators were slowly finding out that military pilots could fly after all. The Sacramento Valley area was a hot bed of crop dusting activity and since we had a lead on a job up there, we loaded our little house trailer that we had parked behind Helen's mother's place and headed out. Helen flew the L-3 and I handled the trailer chores.

We were headed for the Clarksburg Airport, about nine miles south of Sacramento. It was owned and operated by two brothers named Joe and Almer Borges and there was an A&E shop on the field that was run by two partners named John Rakela and Ed Warren. We had yet to meet them but they all turned out to be a really great bunch of guys. The start of John Rakela's last name was pronounced the same as the start of Ruckus so he was always "Ruck" to most everyone. Nicknames were big around the group. Joe's young son answered to Meatball. I'm not sure how that came about.

We had bummed around many airports by this time and it's kind of amazing how many of them seemed to have a resident dumb-dumb. Helen and I had been meeting at certain airports on the way up, as a result she only beat me to Clarksburg by a few minutes. The airport was right up against the Sacramento River levy and as I drove by on the levy road to the driveway, I could look down and see that she was already out of the airplane and talking to a short, somewhat overweight latin-looking guy with one bad eye. I figuratively slapped my forehead with my hand and thought, "Oh, no! She's barely out of the airplane and already cornered by the airport's dumb-dumb!" (Forgive me Almer, wherever you are!)

As it turned out he proved to be Almer Borges and anything but a dumb-dumb — or a pest! Cannonball, (Almer's nickname) Joe, and their wives Corky and Ceil were all a really fine group of people. The brothers were of Portuguese ancestry, and Ceil was Italian, so with Helen being part Cherokee, we all made quite a mix.

Cannon and Joe were both pilots, and in spite of his bad eye, Cannon still flew dusting jobs. Conservative Joe had better sense and confined his flying to ferrying flagmen around, or spotting fields, etc. They currently were operating four J-3 dusters. The Borges brothers not only owned the property that the airport was on but owned and leased a fair amount of farm land in the area and did a sizable amount of farming. They apparently had a somewhat informal agreement that Joe was the head honcho for the farming and Cannonball headed the flying. At least that's the direction their main interests lay in, although Joe's influence was also felt in the flying end. Joe was more conservative in his flying, and approach to business, where Cannon had a slam bang approach to everything. He'd, no doubt, earned his nickname.

To help with the flying business's chores, a non-flying member of the clan named Nelson filled the position of being the brother's right hand man. Nelson was prematurely gray and was often referred to as Grayboy. His main occupation was handling the chores associated with the loading and other ground bound activities. A year or two before we came along they had been operating two prewar Stearmans and one J-3 Cub. As a testimonial to the wisdom of Elwood's admonition to stay away from cropdusting back at Dycer, I'll relate what happened to those three airplanes:

Our ol' Wiley Post!

One of their younger pilots was dusting fields close to the Sierras near Minden, Nevada and made a turn too close to the hills and crashed. Unfortunately the Cub was demolished and he was killed. About 45 minutes after that accident, Curt Metcalf, a well known Sacramento area pilot, was flying one of the old Stearmans right near the Clarksburg Airport when he hit the top of a power pole and crashed in flames. Only the fact that he was thrown free and had his burning clothes put out by tumbling through the dew soaked tomatoes saved him from a longer stay in the hospital. Hardly an hour after that, Cannonball was flying the last Stearman just north of the airport when he elected to let his landing gear brush through the tops of a bush at the end of a pass. This is not an uncommon practice but, as Cannon was unceremoniously reminded, you should first make sure there is nothing hiding in the bush — in this case an old tree stump — and that ended the last of the three planes. Fortunately Cannon survived with only bruises. What had started that morning as a thriving dusting business ended that evening in shambles. Only some borrowed planes from sympathetic competitors would see them through the season and back on their feet.

I suppose that episode, along with an expanding business, helped Cannon decide he would be better off giving up flying the dusters himself. That is how I got on the payroll, using the word payroll loosely, very loosely, as there was no payroll as such. The pilots got a percentage of the gross they brought in, usually 15% to 20%. No work, no pay. In addition to business around Clarksburg we traveled to other parts of the country like Tulelake and Susanville, California and the Reno, Minden, Yerington, and Lovelock areas of Nevada, drumming up work wherever we could.

My getting this new dusting job brought the need for a lot more learning. A much greater variety of dusting material was being used and not all of it flowed at the same rate. In some cases the difference in temperature from a cool dawn to a warm late morning would change the flow rate of the same dust. Then, of course, not all the dusts or crops required the same poundage per acre. In addition to dusts, some insecticides and weed killers were dispersed in liquid form. Grain and alfalfa fields were also seeded by air. Alfalfa was the toughest as there were several types of seed and they were planted at anywhere from one to 30 pounds per acre. Some seeds were also extremely expensive. I remember one job where an error on Almer's part caused him to use an alfalfa job's profit — and several hundred dollars of his own money — just to buy enough seed to finish the 20 acre job.

And the bait! It was the least critical material to apply and paid the least, but it always seemed to be the most problem. Bait came in many forms and densities, although it was always bulky and required many more loads per job. Freshly planted tomatoes required one brand that had a most uneven consistency. It ranged from sticky to free flowing. The only way I could get a halfway even flow was to watch it coming out of the spreader and constantly adjust the hopper door as I flew along. This meant looking backwards with only an occasional quick glance to the front while flying at four or five feet off the ground. It was an extremely trying procedure.

On one occasion I had what I hoped was going to be a one load job as it happened to be about a twelve mile ferry. We stuffed and stuffed and huffed and puffed, but there was no way we could get the last sack of bait in the hopper. In cases like this, you just have to improvise. I climbed in and had the loader lay the large 50 lb. bag on my lap. It made a less than desirable cockpit companion but it was going to eliminate the need for a 25 mile round trip to come back and get it. When I arrived at the field, I located a spot I could land on and threw the bag out on a low pass, put out the load, landed, and threw in the offending sack of bait and finished the job.

In those days there were surplus government issue pilots all over the place, which made it nice for the operators who had finally come to accept them. To insure having plenty of help during the peak rush of the season, they hired more than they really needed. This spread the work pretty thin and most of us had to work at something else in the winter for several years. Usually I worked for Ed and Ruck as an A&E on the field. They did the maintenance work for most of the operators from Walnut Grove to Marysville. Stearman PT-17s and BT-13s were a dime a dozen at this time and most of the operators were having Stearmans converted to dusters with the stronger BT wheels and brakes and P&W 450 HP engines. Although this did not make the ideal cropduster, it was strong, cheap, and carried almost three times as much as the Cub. So the shop was turning out Stearman conversions as fast as it could and as a result, I made the first test flight on many of these newborn dusters. Legally they were limited to 250 pounds of dust at that time, but I have already touched on the cropduster's outlook on regulations. Later on the F.A.A. saw the light and proclaimed that if a duster could get off the ground with it, it must be safe. And the Stearman was strong. I once ran through the statistics and found the Cub had ten times the fatality rate of the Stearman per number of accidents.

Flying the Cub duster was a little like flying combat. Each day you wondered if this was the mission you wouldn't come back from. We had four Cubs but the Borges's were on the Stearman bandwagon also and would soon have one or two. Good thing, because as time went on the Cubs were being used up. Cannon wrecked one in Nevada on one of his rare outings. He had the most blase approach to flying an airplane I ever saw. I think his favorite expression was, "Oh, it'll make it." This time it didn't. At least he wasn't hurt much.

He thought nothing of using the stock 65 hp Cub as a three or four place airplane to show pilots the fields or land to drop off flagmen. This meant landing and taking off with that load in unprepared fields — with one eye! He once came out to pick up two of us pilots who were doing the flagging in a pinch. My practiced eye told me the field he chose was extremely marginal in length. It was deeply plowed at the end and I was sure the Cub would not roll through. I voiced my opinion. "Oh, it'll make it," was his reply. I should have guessed. Reluctantly, I climbed onto my perch which was on the baggage compartment lid behind,the rear seat and pressed my back against the skylight over the head of the crouching pilot under me. This put my head alongside Cannon's, with a good view ahead. About halfway down the field where the last chance to abort was, I said: "It's going to be awfully close, let's give it up." "Naw, It'll make it." Where had I heard that before? At the last possible instant the Cub lifted off, brushed across the tops of the taller clods and staggered into the air.

Another time he flew out alone to flag for me. I was circling in one of the Stearmans when I saw the field he picked to land in. It was freshly mown alfalfa with a rain pipe set up near the end. Even from my viewpoint, I could see it was too short. He overshot his landing and rainpipes flew everywhere, but he just nonchalantly got out and started flagging.

When work was going on, he had the habit of jumping out of the Cub while it was still rolling and letting it end up wherever it stopped. After I'd already finished work one day, I was watching through the picture window of our trailer as the other boys came in when Brash came taxiing in with one of the Stearmans. As I watched him approach the Cub, which had rolled to a stop right at the end of the runway, I thought, "Gee, there's an optical illusion for you, it looks like he's taxiing right into the Cub." It was no optical illusion, he smacked the Cub right in the middle and left it sitting there with a broken back.

Surely all fields of endeavor must have an odd ball or two amongst the more normal cast of characters. Amongst my cropdusting cohorts, the position was filled by a man I'll call "Pete" His physical demeanor was not totally unlike a white "Step-n-Fetch It." He was naive, somewhat gullible, and totally dedicated to flying. Somehow he had persisted his way through college and graduated with a degree in Aeronautical Engineering. This is not to imply that he was brilliant. Oh, he could handle the controls of an airplane fine — it was in the area of judgement and common sense that he fell short of having both wings.

My first exposure to this character came when he was hired as a potential crop duster and was assigned to me as a loader to learn the ropes for a few weeks. In fairness, he was quite likeable and we became good friends. His first effort with me was during the alfalfa season in Yerington, Nevada. My first indication that he had only one oar in the water came early.

We were down to three Cub dusters at this time, and I got a call from Cannon in Reno where he had been ramrodding the jobs with another pilot flying one of the other Cubs. He said the pilot had flown into the trees at the end of the field and crashed (he survived, although badly hurt) and wondered how fast I could finish up in Yerington and get up to Reno to complete the jobs there. I told him I could wind my work up the next morning if I really pushed it.

The fields I had left to do were a considerable distance from the Yerington airport. It was located on the north end of Main Street, a mile long and finished with a very white gravel. My original plan had been to load a five gallon can of gas in the hopper and work off the gravel road near the fields. Pete was to meet me there by truck. But in the interest of expediting the whole thing, I decided to tie the gas can on the wing strut and put Pete in the hopper. By taking off in predawn darkness we could arrive at the job site at the first hint of dawn and get to work. I had no reservations about the load on the Cub. I had done that much and more many times before, except that it was dust in the hopper instead of a person. Keep in mind that Pete was a college graduate with a degree in Aeronautical Engineering.

There were no lights on the Cub, no moon, and not even a hint of daylight when I started the take off. It was black! The only thing that made the dark night takeoff feasible was the brilliant white gravel on the runway. It gleamed enough by starlight that I was able to keep the airplane straight down the runway. Since the night was so dark, and I had a mile of runway ahead of me, I didn't feel the need to make any hot rod short field take off. I raised the tail and let the Cub roll on the main gear for what I felt was a little longer period than necessary. Eventually I eased back on the stick. Instead of the Cub lifting off into the cool heavy air, the tail wheel merely planted itself firmly back on the runway. After one more time of easing back the stick and getting the same results, I began to feel sure that things weren't right. I couldn't see the instruments, but I could tell that the engine was putting out full power. Since there was plenty of runway, I decided to stick with it and see if I could determine just what the problem was.

It seemed like I had used up at least half of the runway when the Cub just lifted and flew off in its normal easy manner. It felt so solid that I suspected that all the takeoff monkey business must have been just a result of my apprehension over the night take off. After all, it had been a long time since I had made a takeoff on a pitch black night. I made a firm turn to the right and headed down the valley.

As soon a I leveled out, the Cub started to sink in spite of the engine still putting out full power. Before I could get too worried she picked up again and real concern didn't set in until it happened a second time. Then there was no doubt about it, something just had to be terribly wrong. I made an immediate vow that if it picked up again, I would head back and try to find that white gravel field.. I could only hope I would be able to see the runway well enough to put the Cub safely back on the ground. It did pick up again, and I immediately whipped a hard 180 around to the left. Again when I leveled out the sinking returned. With nothing but blackness below I mentally crossed my fingers and hoped that it would pass as before. I hoped in vain, she was going on down. I had a three point attitude and full throttle, that was all I could do. The wheels kissed the ground lightly and I immediately cut the switch. I had already locked up the brakes before we touched the ground. Without noise coming from the engine I could hear horses running away from us in all directions during the short roll out. We hadn't hit anything yet, but I was sure we must. Luckily a Cub landing three point with full brakes does not use much of a rollout. We came to a stop and never hit a thing! Even before the Cub stopped rocking, Pete was out of the hopper and hanging on the edge of the cockpit like "Kilroy was here." All I could see were two white eyeballs.

"What happened, what happened? he asked. "I don't know, let me get out and we'll try to find out." I got out, stepped around the front of the Cub and stepped into a lake! While we waited for daylight to survey the situation, I asked him to relate his side of the story to see if it would shed any light on the matter. "Well, it felt good when you started the takeoff. I had the hopper door open and my head out ready to enjoy the flight. When the Cub just rolled and rolled, I thought, boy, we might crash so I ducked down and closed the door." (The hopper on a Cub is where the rear seat used to be and the door is on the top of the fuselage just ahead of the trailing edge of the wing. The door is a little less than the width of the fuselage and about 16" deep. All of this is behind the pilot, out of sight in daylight much less at night.) "When I closed the door, the Cub hopped right off and you made a right turn, so I figured everything was O.K." My hackles were already rising. "Since it felt O.K. I opened the door and we started sinking again..." "You, dumb S.O.B. You mean with your college degree in Aeronautical Engineering, you couldn't tie the door opening with the sinking? You closed it each time you thought we would crash, why did you leave it open the last time?" "Well, when you made that tight 180° turn, I thought it was all my imagination." All I could do was shrug in disgust. The whole morning was wasted just getting the Cub over to a road suitable enough for takeoff.

The effectiveness of that stupid little spoiler was later confirmed. Since the pilot can't see what is going on back there, it was always customary for whoever was loading to slap the side of the fuselage when the load was in and the door closed. You then knew the loader was clear and jammed full throttle and were off. This one time near the end of the load, the loader tripped and accidentally banged the fuselage before he was finished. I was off down the runway and at lift off time, no lift off — just three wheels firmly on the runway. I instantly recognized what had happened from my previous experience but thought I'd continue anyway to see if it could possibly be lifted off. Even with full throttle all the way to the end of the runway and all the way back to the loader, the Cub just would not fly with that door open.

Further indications of Pete's short stay at the brain trough followed in rapid succession. With the ready and cheap supply of surplus aircraft of the day, many a potential entrepreneur bought what they could for resale profit. Pete chose to buy four PT-23 Fairchilds, which at that time was about as poor a choice as you could make. Soon after taking possession of them he put a friend in the front cockpit of one and proceeded to shoot night touch and goes with no lights on the airport or the plane. His flying ability got him through a couple of the landings, but the inevitable is the inevitable. The airplane ended up totaled but, fortunately, the two riders were spared major injuries.

It was not long after this episode that he decided California was "where it was at" and proceeded to ferry one of the PT's out to the coast. On the way, he decided a good way to pass the time was buzzing the deck, which was O.K. since it was wide open desert country. The only problem was that he reached a point where the ground began to rise faster than the old PT could climb. In the ensuing crash he was knocked unconscious and the PT caught fire. His uncanny luck was still holding out, however. A doctor, camping away from it all, was hardly a hundred yards from the crash and rushed over and pulled unconscious Pete out of the wreck and ministered to his relatively minor injuries.

Pete started his cropdusting career back in California. As previously mentioned, his ability to handle an airplane was not suspect but at the end of almost every morning's work, he would come over to me and relate the morning's close calls. "Gee Whiz, I brushed through the crops and bounced off a plow share," or "I didn't see the truck coming and almost hit it." It seemed he had a never ending list of equally hazardous events. I feared for his survival and potential damage to Cannon and Joe's equipment. I was torn between the loyalty of a friend and the welfare of the boss's business. I wound up saying nothing, as not all of his escapades were escaping the two brother's attention.

One particular summer should have shown them Pete's shortcomings with glaring clarity. But first it should be pointed out that they were very happy with his presence because being single, and avidly eager to fly anything, anywhere, anytime, he was always at the airport ready to go at a moment's notice. This feature also wore his welcome with the other pilots mighty thin. Especially since the bosses were more prone to give a job to whomever was handy rather than to think of seniority. And so it was that Pete managed to thoroughly alienate himself from Ed Warren one day. Ed had gone into the hanger to get his helmet for a paid ferry flight in one of the Stearmans, only to come out and see Pete taking off to do the flight for free. Pete had alienated many people many ways. We all liked to buzz the field occasionally but he wasn't content with one pass. He had to do it over and over to the point that even those of us that normally liked watching such things would get sick of it.

But back to this particular summer. Pete was due to take a spray rig back to Montana and for some reason the rig was in the peninsula south of San Francisco. He took off with the Stearman and proceeded to give the airport his series of farewell buzz jobs — which thoroughly irritated all concerned and they turned him in. San Jose was not far off course and he proceeded to give a friend a buzz job there and irate citizens turned him in again. He couldn't pass by our home field without another series of buzz jobs. This included Ed and Ruck's hangar and since Ed was looking for a chance to get back at him for the filched ferry job, he turned him in. He flew on and buzzed people he knew in Minden, Nevada, and that buzz job got him turned in again. He finally arrived at the small town destination in Montana and proceeded to announce his arrival with his buzz jobs and a fairly creditable aerobatic display right over town. Grayboy was to handle loading and ground operations and was already there when he landed. About the only appreciative spectator he could arouse from town was an old guy in an old car that drove up and walked over to Pete. "Whoee! was that you puttin' on that show?" "Yeah, shucks." "Wow, that was the best flying I've ever seen!" "Ah shucks, it wasn't much!" as he kicked the dirt. "No, you were great, you must have been flying a long time." "Ah, heck,"(kicking a little more dirt). "Well, that was great but I'm the sheriff and I have to take you in, you're under arrest."

There was considerable maneuvering and some financial cost on the part of the bosses to eventually gain his release and get him back to work. No matter what Pete did, Cannon and Joe seemed to overlook it. Some things he did weren't too bad, like what happened one morning after his release from jail. He had done quite a number of loads on a field straight out from takeoff. Then the wind shifted and Grayboy moved the loading operation to the other end of the field. On the next takeoff Pete blithely proceeded straight out as usual. But of course it was 180° in the wrong direction! It took 20 minutes of valuable work time before he squared himself away and got back on the right track!

Helen pretending to start our fourth airplane, a Piper Cub J3-65 used as spotter in our dusting business.

Other goofs were more serious. He did a smaller job that required eight or ten passes under power lines near the edge of the field. Later in the morning he was passing by the same field and noticed the farmer standing by it and proceeded to give him a buzz job — and flew right through the wires he had been going under earlier. As if that wasn't bad enough, the sparks caught the tail section on fire! Pete surveyed the situation and decided the fire was progressing slowly enough that he had time to fly on to the strip, which was still some distance away. Even that might not have been too stupid but after making a successful landing at the strip, he neatly taxied up and squeezed the burning airplane into a tight parking spot between two other Stearmans! Cannonball was there witnessing the unbelievable. He jumped up and down and started waving and yelling trying to get Pete to get the plane away from the others, but to no avail. The only recourse left was to rush up and kick and tear a fire-break right behind the cockpit with feet and bare hands — then stand and watch the tail section burn.

By rounding up another set of tail feathers, it was finally decided (with Pete's urging) that the plane could be ferried back to California for full repairs. Which he did. The only problem was that it was done without a ferry permit and he was turned in to the F.A.A. while stopping for gas at one or two of the more formal airports. It just added to the bundle of citations he had already gathered. Somehow, being quite likeable to most people, he managed to resolve his differences with the F.A.A. with a minimum of fuss.

His naive gullibility made him fair game for practical jokes on occasion, some a bit on the cruel side. During one stay at a Reno hotel by several of us, we couldn't help but notice that the hotel maid was pretty crusty and quite homely — perhaps even the homeliest woman in the world. Pete was extremely shy and backward around women and would instantly fall in the "Ah, shucks, heck, kick the dirt" mode at the mere presence of a female. One of the more heartless of our group decided to counter the dullness of the stay by telling the maid how Pete really had the hots for her but was too bashful to do anything about it. The ensuing maneuvering by the male starved maid and Pete, who wanted no part of the situation was pretty humorous.

A time finally came when I thought my worry over loyalties was resolved. Pete, among other things, was an avid skydiver. On one of his Sunday airshow outings, he broke his leg and I thought, "Good that takes care of this season." But a few days later there was Joe helping him into the Stearman, cast and all! The final resolution came with the Korean War. Having missed the action in WWII, Pete decided to volunteer his way into action with F-86 Sabre Jets. Somehow that didn't last long. The only story he came back with was about bad weather while leading a flight of four one day. He called for radar to his home field and was told it was no go, zero-zero. He then asked for a vector to the only other alternate and received this reply: "Sorry, the alternate is also zero-zero, there's nothing I can do for you, thank you and good day."

With flying time left measurable only in minutes, the situation was not good. Amazingly a very tiny break in the clouds showed up soon after and with a violent maneuver Pete managed to take advantage of it and somehow found a place to land. Unfortunately, his element leader and wingman were unable to follow the maneuver and were never seen again. Pete was then put on the carpet for failure to take proper care of his flight! Although I found fault with much of Pete's judgment, I felt the Air Force was completely in error on this point.

It seemed obvious from the beginning that Pete's luck could not continually save him from his bad judgment, although time and time again he did come through. Near the end of his service tour, he somehow managed to overcome his awkwardness with women and finally got married. He quickly fathered two children and appeared to be on his way to a happy family life. It was extremely unfortunate that his luck finally ran out on him during a vacation flight over the Florida Everglades in bad weather. An accident took his life and his family as well. It was a hard way for the inevitable to catch up with him.

My first forced landing in the Sacramento area came one morning just south of Clarksburg near the Sacramento River. It was the first

Helen surveys the damage to the Borges Brothers' Cub duster.

load of the morning and the engine lunched right in the middle of a pass. I just barely cleared a ditch and made a landing in the next field. Since this area was much more populated than my previous forced landings, my first worry was about controlling the crowd that would surely materialize. Not only did no crowd show up, but none of the crew did either. I finally had to walk a mile and a half to a phone to get help.

Three more forced landings happened within a mile of the first one in the next few days. The last one came while I was making my last pass on a small tomato field right next to the river. I was just about to go under a low power line when the engine let go one more time. This would be the fourth forced landing in ten days. This time the engine let go with a much bigger bang and flames licked back toward me from all round the firewall. As usual it was the first load of the morning and still very dark, so the flames looked awfully fierce. I had to continue under the wires before I could pull up. The river curved around in front of me which precluded going straight ahead so I pulled up to the right, quickly using up what extra speed I had. This left me on a close approach to a nice freshly mown alfalfa field. The only trouble was the hay was stacked in big randomly spaced haystacks — dry haystacks. It was the only time I had ever seen hay stacked that way in the Clarksburg area. It was doubtful that I could miss them all and I had to be coming in with a flaming engine!

Of all the ways to die, I fear burning most of all. I decided to unbuckle, get out on the wheel and reach in and fly with my left hand. (How I knew I could do this will be explained in a moment) My plan was to wait until it was near touchdown and then jump. I figured a 40 MPH roll in the hayfield was better than burning. I was trying to get out as fast as possible, but the little half door stuck and it took a couple of shoulder bashes to get it open. On top of that the seat belt was not the quick release type and after trying twice I still didn't get it released all the way and was jerked back into the seat. By the time I did get it loose, it was too late to get out. I just had to stay put and try to miss the hay stacks.

It was just as well because right after touchdown the fire went out and I was able to squeak a zig-zag course through the haystacks. Ah, another successful forced landing — almost. A ditch suddenly loomed up and I tried to ground loop to the right to miss it, but I couldn't quite pull it off. I went into the ditch at a 45° angle at about ten MPH. Both lower longerons bent in two places, the gear suffered along with a wing strut or two, and my head rattled between the two bars in the cockpit. Except for a knot on the side of my head, everything was O.K. with me, but the Cub was going to need a little repair.

A decision was finally made that the old A-80 continental that had been powering the Cub was having to be pushed too hard for the up to 600 pound loads we had been carrying. It was replaced with a later model C-85 Continental. I was surprised when the C-85 didn't seem quite as powerful as the old A-80. No wonder the hopped up A-65 (ne A-80) gave up the ghost so often.

The reason I knew I could fly the Cub while standing on the wheel was because of some of the stresses and pressures of being a crop-dusting pilot. It can lead to an occasional unorthodox approach to some of life's little problems. Take the long boring cross-country I was taking in the J-3 Cub duster one day over the barren stretches of Nevada country-side, for instance. The J-3 is not a fast airplane, it is not at all unusual to have the faster car-bound traveler pass you by.

I was struck with the urgent need to take a leak but time was of the essence and I didn't want to waste any by landing on the road or occasional dry lake. By standing with one foot wedged between the wheel and gear tubing, and reaching in with my left arm to fly, I figured I had it made to pee in open air. Being of an analytical mind, I figured facing toward the wing tip would let the shear wind whisk things away with no problem. Somewhere my calculations went awry. I then tried facing the tail, a position I had earlier discarded knowing about the effects of turbulence and found my calculations to be correct. The only alternative left was facing forward, a position I figured to be the worst of all originally. I was right again! I gave up, climbed back in and peed on the floor! I figured that it couldn't possibly be any worse on the plane than some of the fertilizers we dispensed.

The other of the last two Cub dusters was being flown south of Sacramento one morning by another pilot. Something happened on one of his takeoffs and he crashed in flames. It was a tragic end for them both.

The bosses had been gathering 450 Stearmans, but the Stearmans were not as maneuverable as the Cub and there were many, many small fields with wire and tree obstructions along the Sacramento river. Of course they were far less lucrative for the pilots — and far more dangerous than the big open fields. By now I had become senior pilot in the organization and could have moved into Stearmans. However, since I had become quite handy with the Cub, I was offered 40% of the gross, an unheard of percentage for a pilot — if I would take all the bad fields. I had already discovered that I could keep up with the average Stearman pilot in pounds of dust put out in a morning, even doing the bad fields, so I quickly agreed. Although the Cub was much slower than the Stearman, I could make the turns about three times as fast and often load closer to the job than the Stearman. It all helped.

Heading out one morning in the first dim light, the 85 quit again. This time it was a broken fuel line to the injector. For a change I had about 200′ of altitude but it was pretty dark and I had 500 pounds in the hopper. Luckily there appeared to be a good level field on my right, so I took it. There wasn't much else to choose from anyway. The field turned out to be smooth but deeply plowed and I could feel the wheels grab as soon as I landed. I quickly induced as much of a ground loop as possible to put the load to the side so she wouldn't go over. It worked, I doubt if the Cub had gone 50′ before she sank nearly to her belly. It took several hours and six people just to manhandle it over to a suitable takeoff point.

Again I admit to some haminess. The old airshow days had made their mark. I took the dusters into spots, under things, and between things few pilots would attempt. It was not at all uncommon to intentionally brush a leafy tree or bush with a wing tip or wheel just to get into a tight spot, or under a low wire. The bosses didn't mind and farmers often requested me for a job over the other pilots. I'm not sure whether it was for the entertainment — or for what they thought might be my inevitable end. Only once did I nearly oblige. It was on a field in Nevada next to a road. There was a line of pretty good sized Weeping Willows hanging well over the field. Of course it isn't all ham. It's important to cover as much of the field with the material as possible. In this case to adequately get the first pass in, I had to get the right wing under the tree limbs which were quite low. This meant doing a forward slip the entire pass to keep the wing low enough to get under the trees. About midway there was a tree with a little extra leafyness hanging too low to get under but I decided it looked harmless enough to brush through. What should have been a hardly felt touch, turned out to be an audible whump and jerk. I went back by truck after the work was done to see what there was about some little flimsy Willow tree leaves that could put a hole in my wing. It turned out to be a very husky branch that apparently had nearly reached the ground and had been cut off leaving the thick sawed off end neatly hidden by the leaves.

Well... O.K., maybe twice I nearly obliged. The other time was while doing a very small field near Reno. The altitude of Nevada was tough on the Cub's performance, I needed all the horsepower available. Just as I dove down past the trees and leveled out on the field, the engine started to ice. With my still youthful fast reflexes, I quickly yanked out the carb heat — and I do mean yanked out. I ended up with the knob and part of the rod in my hands. There was just enough stub left sticking out of the panel to push back in. It was then that I made a cardinal mistake. I spent too much time in the cockpit trying to decide whether to push the stub back in and get back all of my horsepower and risk icing again, or play it safe by leaving carburetor heat on and suffer with the reduced power.

I chose the latter and looked up just in time to see the barbed wire fence at the end of the field right at my nose. Only immediate stick action would allow me to avoid running full tilt into the fence — and quick stick action I took, much too quick — it resulted in a climbing half snap roll. There was only time for a short moment of amusement as I looked up (down) and saw the farmer and his family, who had been watching from the corner of the field, start to run towards the spot they thought I would crash in. Fortunately, my familiarity with aerobatics permitted a close, but safe recovery. Well, I never said I didn't make any mistakes!

There may even have been an occasional lack of patience. I tend to be pretty short fused if I feel my toes are being stepped on unnecessarily. One morning I was approaching a tomato field with a load of pure sulphur, a totally non-toxic, but very irritating material to humans. It is extremely irritating to the eyes and adds to the discomfort of sweaty necks considerably. As a crop dusting pilot, you become aware of this with the very first load. I spent many hours modifying goggles in an effort to get the best seal possible. Fortunately, the body develops some tolerance to sulphur so that by the end of the season, it is only miserable rather than intolerable. As I neared the field, I noted a gang of wet-backs working right at the edge of the adjacent field. I carefully put out a small dab of sulphur on the far side of the field to confirm that the slight wind drift would carry the sulphur away from them. I then started my passes right next to the workers to be sure and work away from them before the wind shifted.

They were all bent over working away until I got too close to dodge. Then, just as if their commander had yelled "fire!", they all threw whatever they could lay their hands on — shovels, hoes, rocks, or whatever. It made about as fine a coordinated attack as I've ever seen. Fortunately the missiles all fell about five feet short, but was I angry! I pulled almost straight up, did a wing over, and pulled out right over their heads pulling about two G's. I popped the hopper door wide open which doused them with a whopping load of sulphur. I can assure you they were about to spend the most miserable day in the hot sun of their lives.

Back in Yerington, Nevada I woke up from a quick nap in a car one morning. I was due to go back to Sacramento in a day or so and none too soon since I was dead tired and ready for home. Taking a quick nap in a car was not all that unusual — but to wake up in the back seat of a strange car certainly was. Even more unusual was recognizing the driver of the car as a farmer from the Clarksburg area named Harold. "Harold! What in the world are you doing in Nevada?" "We're not in Nevada, we're in California!" He seemed quite tense. "Harold, I'm in Nevada, and if you're with me, you're in Nevada!" "We're not in Nevada, dammit! You've had a plane crash, and we're taking you to the hospital!" He was too pale and thin lipped to be kidding, but I replied, "Aw, Harold, you know I never crash." Peeking out the window from my slumped position did reveal trees and a levy that looked very much like the Sacramento River area. As I tried to sit up straight another friend at my side put his hand on my shoulder. "You better lie still, you might have some serious injuries." I certainly didn't feel any pain and was 100% alert but I started feeling around for injuries just to humor them. They were much too grim to be kidding.

Well, my pants turned out to be torn, and my gosh — my face was covered with semi-dried blood. In fact my head was a mass of cuts and abrasions. What the heck was going on! My knees were even stiffening a bit and had some deep gashes on them showing through the rips in my jeans. I couldn't detect anything seriously wrong though and started scooting up to a full sitting position. The farmer at my side still insisted that I should stay still. "No, I'm O.K." "Are you sure?" "Sure, I don't feel too bad at all." I took a better look outside. Sure enough we were on the Sacramento River road! How in the world could that be? The last memory I had before the "catnap" was being in Yerington, Nevada. By the time we got to the hospital I was beginning to hurt all over, but there was still no sign of a serious injury. Also my last memory of being in Nevada had advanced a couple more hours. As I was being helped with my slow hobble into the hospital Cannonball and Grayboy showed up. "Boy, am I dingy! What happened anyway?" I asked. "Well, you totaled the Cub, don't you remember it at all?" "No, as far as I know I'm still in Nevada!" I knew by then, of course, that I wasn't. "How did it happen?" "There were several eye witnesses. One was sure the engine failed, another one was just as sure you stalled, a couple more weren't sure so take your pick."

I couldn't buy any of the accounts. My going in from a stall was completely out of the question. An engine failure might have caused light damage in a forced landing but even that couldn't have been the cause since I had ended up in the middle of an open field with the airplane totaled. Helen was in the hospital recuperating from giving birth to our son. I called her on the phone to give her the news. Her voice on the line and the trauma finally caught up with me. I broke down and it took a moment or two before I could talk. The doctor didn't find any major injuries so after his check-up, I hobbled back out to the car under increasing discomfort. I had a couple of broken ribs and a chipped elbow and too many cuts and abrasions to count. My back would never be quite the same. "Let's go look at the wreck. I want to stop by the house and pick up my camera on the way and change my pants." I had hit so hard that the seat of my jeans was shattered like a pane of glass.

Remains of Cub, close to being totalled, the result of a broken wing strut.

As we drove along, a couple more hours of memory came back. As it turned out, exactly three days had been wiped out by the wreck, but it was slowly coming back. We got to this nice big field and there, near the middle, sat the cub. Indeed, it was close to being totaled. only the last foot or so of the fuselage and the tail surfaces had survived. The wreck looked unsurvivable. The engine was pushed back into the gas tank on the front side and the back side of the tank had two nice big indentations that matched the shape of my knees. There no longer was room for legs in the cockpit and not much room for the rest of the body. There were three closely spaced gouges in the field, including the one the plane was in, indicating very little forward motion when it hit. With no memory of the crash, there wasn't a clue as to what might have happened. More memory continued coming back, however. I had gotten back to Clarksburg the day before and I now remembered making preparations to leave Yerington that morning. I wanted to rest a while and went home until the time arrived for the F.A.A. to come and take a look. We were good friends by then from all of the dealings over the years.

Soon Rysaavy, our F.A.A. man, arrived and we drove out to look at the wreck. By then my memory was better still. I remembered the flight home from Nevada the day before and loading up for the job that morning. As we drove along my memory continued advancing. I soon remembered working the field.

Then it all came back! I had been putting out bait which is light and coarse textured. To get a good spread we flew the passes quite high and the turns were even higher. I was in the middle of one of my usual tight turns when suddenly there was a "thump" that I felt more than I heard. Something had broken! A few days earlier, one of our mechanics had come over to warn me that some of the tubing back by the tail wheel area was pretty well gone. The tail had been shaking more than I thought it should in rough air. I thought it would last until license time, however, which was just a few days away. With the thump, and the ensuing uncontrolled roll into the turn, I just knew I was wrong. I didn't even bother looking since I was sure that the tail had given way.

With structural failure, reduced speed is a must so I snapped the throttle back — so much for the engine failure eyewitness! At the same time, I gave full opposite control to the roll and it stopped the airplane absolutely vertical. The nose then fell through and pointed straight down — so much for the eyewitness's stall! The only thing I had left to try was full throttle. I was running out of altitude and other ideas fast. Full throttle brought the nose up in a knife edge flare. That was good, but at too high an altitude. I let the nose fall back down in order to flare for a knife edge landing on the wing tip — not normally an acceptable landing, but better than going straight in. At the right moment I gave full throttle again but it didn't seem to give the same response that it did before and the ground was nearly here. I grabbed an overhead tube with my left hand and held the control stick over with my right and ducked my head with the last thought I can remember: "Boy, this is sure a steep angle to hit the ground." The memory of hitting the ground has never come back.

As the last gap filled in, I exclaimed to Rysaavy: "Wow, I remember now, I remember! something broke." "Are you sure?" "Absolutely, there was a thump. Something broke!" But there was that undamaged tail sticking up like a banner. The cause would have to be found elsewhere. As we poured over the wreck for clues, we noticed all of the wing strut forks were bent and broken except one — it was broken, but straight. A close look at the broken ends revealed rust a third of the diameter and the unmistakable crystalline look of a fatigue break on the rest of it. We had found the culprit. only the aileron cable running up the strut had kept the wing from folding all the way back.

The next day I was still looking a mess when I hobbled into the hospital to visit Helen. As I entered, a passing nurse took one look at me, grabbed me by the arm and started to steer me toward emergency. "No, no! I"m just visiting my wife here, but thanks anyway." She gave me a look like she wasn't real sure I had all my marbles, but let me go anyway.

It was back to work in the Stearman the next day, under protest from the old bod. Every muscle and joint I owned was sore. It was several days before I was back to near normal. So went the last of the Cubs. Three times I had taken phone lines down with them. In rural areas the phone companies have a habit of tacking them in the most surprising places, and from the air at 60 to 80 MPH, in dim light, they can be difficult to see. Especially if your eyes are already suffering from sulphur. Of course, the absolute safe way to approach dusting would be to scout each field ahead of time, on foot, and determine where all the hazards are and memorize them. However, if you did this, you would get little rest — or work done. Most of the fields I did, I never saw until I approached them with the material — either led there by spotter plane or by hand drawn maps, or sometimes just a description over the phone.

The old Cub had been a lot of fun. I once came back to the Clarksburg Airport to find that one of our frequent north winds had come up, it was really ripping. Whenever I hit those conditions, I couldn't resist seeing how slow the Cub could fly. This day it would not only fly slow, but backwards. I struggled upwind to the bridge about a mile north and flew backwards to the airport. Then I played around for a few minutes like a helicopter going straight up and down. I finally decided I might be able to land with zero roll. It would be tricky because there often is a layer of relatively dead air near the surface. I decided to give it a go anyway and it worked, I had made a zero roll landing. of course in that gale, it was pretty spooky taxiing over to the shelter of the hangar. Some of us never grow up. In a way I would miss the old Cub duster.

The only time I ever saw the boss even remotely irritated over our flying was on a slack day when three of us were just shooting the breeze around the gas pit. Another pilot took his Stearman up for a test hop after a repair of some kind. One of the two guys couldn't stand the temptation and sneaked out with another Stearman to bounce him. He was shortly followed by the last two of us and we soon had a four plane dogfight going right over the field when the boss drove up. Upon landing, we were greeted by a slightly grumpy, "What'er you guys doin'? That's my gas yer burnin' up!"

Most of us pilots were down to earth and ready for good laugh anytime — or an on the spot competition. One of the operators in the Marysville area was quite a huge fellow, well over six feet and built like a fullback. He was traveling by truck from one work site to another when he was struck by one of nature's regular calls. Since he was passing near his home, he dropped in and rushed for the bathroom, only to find his very young son sitting on the toilet in his training seat. When he looked out a back window he saw his wife talking to a neighbor lady over the back fence. He went back to the bathroom, set his son aside, sat down and left a man sized deposit. He then carefully sat his son back on the toilet and hid in the closet. His wife soon returned and took one look at what she thought her son had passed and yelled in disbelief! She ran out and brought the neighbor lady back in to see the unbelievable! History does not record how the by-now embarrassed husband resolved the mystery for them.

One of the more hilarious competitions occurred south of us involving Vic Spezia. It was quite a windy day in the Delta, as it often is, and a group that included a couple of pilots and farmers were sitting around Vic's strip which backed up to one of the many sloughs in the area. one of the farmers owned a J-3 Cub that was sitting on the line and talk soon got

Me at work in a Stearman duster.

Line up of Stearman crop dusters owned by Borges Bros. at peak of business. Clarksburgh Airport.

around to discussing how short the Cub could be taken off in the wind — and who could do it the shortest. Vic was first up at a make shift starting line to give it his best shot. The Cub owner was to mark the lift off point by standing a little ahead and to the right of the Cub.

Vic locked the brakes, brought the engine to full power and just as he released the brakes, a gust of wind picked up the left wing sending him careening towards the Cub owner. The Cub owner immediately turned tail and ran — but not fast enough! The low right wing caught him right behind the knees and flipped him backwards over the wing. Vic, still desperately trying to save the situation, took off in a somewhat out of control climbing right turn where the gusty wind put the "coup de grace" to the whole thing by blowing him right through the wires. That caused him to drop right into the middle of the slough. Fortunately a boater was on the spot to pick him up and Vic wasn't hurt a bit. It seems a shame the takeoff distance wasn't recorded!

I managed to pick up $5.00 on a Cub takeoff bet one day. The wind was blowing pretty good and several of us were standing around the line slinging the bull. We were all bragging about what we could do with our airplanes. I mentioned that I could start a takeoff in the Cub duster from its tie down spot, aim towards the gas pit, and be off before I got to it. I was immediately challenged by this Texas boy in that very positive way many young Texans have. I countered that not only could I get it off by then, I would clear the pit by 50 feet. He immediately pulled out a $5.00 bill, "This says you can't do it!" It was an easy bet, and really unfair, as I had done it many times before. The distance was very short, probably no more than 200 feet. No, it may not have been fair, but when did a loud Californian ever pass up a chance to put down a loud Texan?

I fired up the Cub with the Texan standing near the pit and let her rip. I purposely teased him by holding the Cub down until the last second. When I was about 50 feet away, I looked at the Texan and he was grinning a winner' s grin. I pulled the Cub into the very brisk breeze which let it climb virtually straight up. I looked down to see the Texan stomp off towards the office. I landed and went over to him and asked. "Are you satisfied that I had 50 feet over the pit?" He handed over $5.00 and said, "Hell, you had 500 feet!" I kept the fiver.

A good share of my forced landings came in the Stearmans. One came while I was spraying a field that had a wide slough and levy at one end. Just as I was approaching the levy, the engine quickly went to idle, in spite of my firm grip on the throttle. It turned out that one of the mechanics had failed to safety a bolt on the carburetor linkage and true to Murphy — if it can, it will. The engine might just as well have quit for all the good it was now. I was able to make it over the slough and was then faced with the decision of side slipping like mad to try to make the near short field and possibly overshooting to the road and inevitable ditch, or go for maximum glide and hope that I could reach what appeared to be a beautiful field on the other side of the road. There was also a very low phone line along the road. Either choice would be marginal but I chose the latter. I was already in three point attitude when I skimmed over the road and under the phone lines that were so low that I wasn't sure they wouldn't snag the gas cap on the center section. I had already shut off the switch and it greased on three point about 20′ past the road. My relief at another successful forced landing unraveled when I felt the tires grabbing soft dirt one more time. The field had been deep-plowed, dragged smooth and was as soft as cotton. I didn't try to ground loop it like I had the Cub, she was nosing up too fast.

As it went up on its nose, I gave a loudly voiced opinion of the field's ancestry. It came to a stop straight up on its nose and one wing tip. Amazingly the field was so soft there was no damage of any kind. While we were getting things in order a little old lady came out of the nearby house, and said; "That's a close call," she repeated it a few times and added, "I even heard you scream from inside the house!" What!? I had never screamed in an emergency in my life. It finally came out, only after some prodding, that what she had heard was my very loud, "You son of a b____!" She was just too much of a lady to repeat it!

Soft, plowed fields tended to make for short, unconventional landings. This one, as described in the text, didn't even hurt the prop!

The mechanics replaced the bolt while I scouted the nearby road. It appeared to be the only suitable takeoff spot around. I paced off the distance from the left shoulder of the road to the phone line poles on the right. By placing the left wheel as far onto the shoulder as possible there would be less than two feet clearance between the right wing tip and the telephone poles. Eventually it was time to try the takeoff. It was close, but we made it. Keep in mind that a 450 Stearman is about as blind as a P-51 or P-47 in the three point position. However, In our operation, loading off narrow boondocks roads as well as other undesirable spots was a frequent requirement.

I remember going out for coffee with Cannonball one morning after work. While we were out he decided to show me the field we'd be loading off of the next morning. When we stopped after a bit I asked: "How much farther to the field?" "This is it." "This is it?" We were looking at a short freshly cut alfalfa field that my practiced eye could see was shorter than the landing roll of a Stearman — and it had big ditches all around it. I said so. "Oh, it'll make it." I might have guessed his response! "Well, it's your airplane."

The next morning it looked even smaller from the air but there was the dust, 12 loads of it, ready at the end of the field. Takeoff would be less of a problem because the 450 Stearman had the unique ability of taking off with up to 900 pounds in a little less distance than it took to land — providing you locked the brakes and ran full power before you started the take off roll. But first I had to successfully land the thing. I carefully planted it three point just inside the near ditch and with maximum braking watched the far ditch rapidly approach. If I had just gone straight ahead I still would have wound up in the ditch; but by bringing into play my good old controlled ground loop, I was able to dissipate the last bit of roll and stay out of it. This also conveniently placed me right in position over the sacks for loading. This had to be repeated 12 times with no margin for error.

The Stearman was a good, strong airplane and the P&W 985 was a very good engine. However, many of the engines we had were old and tired and the dusting operation certainly abused them so engine failures were far from unheard of. Most were at very low altitudes, often in the early morning darkness, and most inconvenient. We had one particularly tired engine for a while. I wasn't too thrilled about flying it at any time, much less on a long cross country. But, some work came up at Susanville, California, and Lovelock, Nevada, and Cannon wanted me to take this dog to get it done. I objected, voicing my opinion of the engine. I should have known better. "Oh, it'll make it."

So off I went. You spam can drivers with your electronic aids and juke box displays may find it a little hard to believe that these Stearman cross countries were made with no maps in the windy cockpit and no compass. In fact, if it was well into the season, the only instrument still working most of the time would be the oil pressure gauge. Of course, this dictated flying under IFR almost exclusively — that is, I Follow Roads.

After getting the Susanville work done and while working the Lovelock jobs, the old 985 decided to pack it in — thoroughly. The master rod let go and of course this just about ruins everything else but the accessories. Naturally it had to be right on the deck over very modest sized fields with big levy-ditch borders. There was very little to choose from, but I made it all right. The farmer turned out to be an agreeable soul. "No problem, no hurry. You've got clear 'til tomorrow afternoon to get it out. That's when I get my water and I'm flooding the field." Oh thanks a lot!

It meant many wakeful hours for the mechanics to get an engine here from California and installed. Of course, it also meant I'd have to test fly their hasty handiwork off a very short field towards a very tall levy. Well, they made it, and so did I. We beat the water by 30 minutes and cleared the levy at the end of the field by inches.

When I think of all the forced landing situations, the ones I faced on takeoff were the most unnerving. You're in what had been a two place airplane loaded with the equivalent of eight or nine full sized passengers and still flying at minimum airspeed with no extra altitude — all combined with no visibility over the nose. I've had at least four at the Clarksburg Airport. The most unnerving one came when I was heading out to the south where the field selection is less than desirable. All of the fields had ditch borders — some as deep as eight feet. Just after commitment the engine blew a cylinder head. This did not cause complete failure but, with a full load, it assured me of a quick landing. The accompanying oil and smoke from the engine was further unnerving.

The length of the fields, spacing, and lack of altitude presented only one possible landing spot. I tried for it but even with full throttle I couldn't quite reach it. She greased on three point, but just short of the near ditch. I immediately hit the switch and braced for what had to be a bad one. But there was no crash. When the wheels reached the ditch I felt nothing more than the usual bumpy roll out on an unprepared field. Hard braking brought me to a stop just short of the ditch on the far side. Unbelieving, I walked back to the ditch I'd gone across to see what miracle had let me off one more time. It turned out that the field I ended up in had been recently land-planed and there was about a foot difference in height of the ditch sides — in my favor. The wheels had skimmed from the high side to the low side with scarcely a murmur.

Forced landings seemed to come from all causes — some particularly unexpected. I was on a long ferry once with a full load of liquid. It was an unusually long ferry of about 12 miles. For one or two loads it isn't practical to shift loading points. I was cruising comfortably at about 500′ following Joe in the spotter plane south of the Aerojet Plant when I looked down for one of my infrequent instrument checks just in time to see the oil pressure needle drop like a wet noodle. I wanted to save the engine, and get down while I still had the use of it, so I picked out the best part of a back road to land on. The fields were totally out of the question.

The road turned out to be like a roller coaster — and almost as narrow. As usual I cut the switch at the three point touchdown and

A typical day at the office.

rolled over the hilly road. After braking to a stop I climbed on top of the center section and still could not see over the hills about 100 yards in either direction. Of course this completed only half the saga. With the engine replaced I still faced a takeoff on the same road and still fully loaded. It was all somewhat challenging. The oil pressure drop turned out to be caused by a mechanic overlooking a rag stuffed in the engine on assembly. At least no harm was done and the engine was still O.K.

So, like I said, the P&W is a good engine basically, but we did put them to hard use. I've had other forced landings caused by a thrown master rod, two thrown link rods and four blown cylinder heads. The blown cylinder heads were four of five that blew on one engine alone at different times. one of them blew with a bang just as I shut the switch off at the gas pit. Gads, I spent many an hour at that gas pit with the ding of the bell marking every gallon. Many years of engines and wind in my ears would eventually still that bell. I finally couldn't hear it at all if I was more than a few feet away.

None of the forced landings in the foregoing were my fault, and with a couple of exceptions they were executed with no damage. I would be remiss, however, if I didn't relate the one forced landing that was my fault. Nobody's perfect. Let me first take some of the edge off my guilt by pointing out that cropdusting is a pressure business. The hours suitable for working are limited and any advantage possible must be taken of the time when it is suitable. Rush, rush, rush is the name of the game. It is a very physically and mentally fatiguing venture, especially in an open cockpit Stearman. I have been so tired near the end of a long day that I have actually dozed off on the way to a field only to be brought awake by my chin hitting my chest — this in a roaring 450 Stearman with about 120 MPH of wind going by my head. I might add that I don't drink or smoke and was always in top physical shape. Yes, I know, it was time to stop, and I did.

When operating the business myself one morning, I dropped a heavy spreader and caught my left thumb between it and the edge of a five gallon can. It completely shattered the bone. The working day had just begun, though, so I just had to get on with it. It was several weeks before I had a doctor look at it. "Well, I can't do anything with it now," and he sent me the bill.

An undignified landing for "my" Stearman, due to over irrigation of surrounding fields flooding the runway.

I broke my ankle on a motorcycle hill climb in the middle of one season. My oldest son Lan built me a flexible brace with a hook at the foot of the cast to latch onto the toe brake. With my loaders getting me up to the cockpit on a forklift I was able to get on with the season. After a few weeks I was able to forego the forklift and worm my own way into the cockpit.

The business was not conducive to showing mercy on myself or others. For instance Helen picked up the loaders one morning and I was already warming up the plane when they returned. After a little delay one of the loaders came running up white faced. "Helen just bashed her finger in the car door and fainted." "Is she comfortable?" "Yes, she's lying on the car seat." "Well, O.K., let's get to work." After a load or two, Helen wandered over, not much worse for wear, at least there were no broken bones.

If I have evoked enough suitable feeling of empathy, I'll get on with my stupidity of the forced landing. It was near the end of one of those long days of spraying and, as usual, I had stretched the gas to the limit. As I approached for the landing, I was already mentally reviewing the procedure: load up, get the loader on the wing walk, taxi over to the gas pit, gas up, and go again.

But, as I taxied up to the loading area I could see that Joe was there pacing back and forth with considerable arm waving. It seems a minute amount of plant killing 2-4-D had found its way into a very expensive load of insecticide that I was about to spray on tomatoes. He was agonizing over whether to throw the valuable load away and eat the expense, or go ahead and use it with the very small possibility of damaging the farmer's tomatoes. I soon joined the arm waving and pacing. Finally, after about five minutes of agonizing, we decided the farmer's interests came first and we chucked that expensive load. Eager to make up for lost time I hurriedly took off with the fresh load. With all the confusion the need for gas had completely slipped my mind as well as Joe's and the loader's too. They normally were always cognizant of the fuel situation and quick with reminders.

I was in the middle of a steep turn right over the Sacramento River when the engine quit. This was going to be a close one. With a heavily loaded airplane I would have to finish the turn and hope to be able to clear the trees and levy. I did clear them, but there was a set of power lines too high to go over and so close to the levy that getting under them would tax my ability. I just made it. Now all that was left was to see what a landing in ten foot tall safflower was like. My mind was still occupied with the vision of that several hundred dollar load going down the drain so I was mentally cussing that P&W for letting me down one more time.

I didn't have long to wait for the answer on the safflower. As soon as the wheels touched, the safflower clamped on like a vise and flipped me over at full landing speed. I squeezed down in the cockpit with the thought that this was going to be a bad one. A few weeks earlier I had returned to Clarksburg after the first load of the day. It was still quite dark and I didn't know that the end of the runway was flooded from Joe's over exuberant irrigation of the adjacent tomatoes the night before — neither did he. I landed right in the middle of it. The water wasn't real deep, but it caused enough drag to flip me near the end of the roll out. Even as slow as I was going, it still hit upside down with an awful bang. My shoulders were sore from the shoulder straps for a couple of days. So I was expecting the worst at the speed it was going over in the safflower, but I got the best. That safflower was so tall and thick it made the whole thing very cushiony.

I had been so distracted by the loading error that I was still cussing that blankety P&W

Our Waco YKS-7

for quitting again as I scrambled out from under the plane and clawed my way out of the prickly safflower to the top of the levy. Then it finally dawned on me! I had run out of gas! It was unforgivable! The mortification and feeling of guilt is hard to describe. There is just no one else you can blame. Responsibility for monitoring the fuel belongs to the pilot only. With a gas gauge right under your nose there is no excuse for running out. I plead Nolo Contendere.

On top of that, I was still faced with the difficult task of getting the plane back in the blue once we got it out of the safflower. But first a slightly bent prop and rudder had to be replaced and the plane hauled to the top of the levy via the adjoining driveway. With the help of some strong backs and a tractor we got it done. That left me on a narrow curving levy road to take off from. We took down some interfering mail boxes and I was ready to go — after adding some gasoline, of course! Several cameras were ready, just in case, but we made it anyway, perhaps to their disappointment.

I wouldn't want to unnerve neophyte pilots, but not all of my forced landings have been in hard-pressed cropdusters. I have always enjoyed different kinds of airplanes, and changed our personal airplane quite frequently. So it was that when a chance to trade a Luscombe 8-A we had for a YKS Waco Cabin straight across came along, I took it. It was a fine old airplane powered by a "Shaky Jake", a 225 HP Jacobs, that is. It was a clean, smooth running engine actually and I was eager to get Helen checked out in it, all 100 pounds of her. She had kept up with the flying, and took her F.A.A. private license check ride while five months along with our first born. At this time the test still included having to do tail spins.

On her very first takeoff in the Waco the engine lunched right off the end of Clarksburg Airport. I wasn't too worried, I simply took over and landed in a field where I'd had plenty of practice. In the past year I had used it twice with engine failures in the Stearman. Fortunately I found a complete, but derelict, Bamboo Bomber with two low time Jakes that I was able to buy for $150.00. Those were the days! We were soon in the air again. Helen checked out in the Waco with an absolute minimum of problems, and flew the old bird all over California.

We owned numerous airplanes after the Aeronca L-3, but the L-3 was a fine old machine and trouble free. Well, almost. While teaching an eager lady student how forced landings could be safely done in seemingly poor areas, I landed on a back levy with a one lane dirt road. I had to shut it off to turn it around, not a bad trick in itself, since the levy was considerably narrower than the L-3 was long. On restart, the rather old harness decided to show its age and the engine could muster no more than about 70% of its 65 HP. I tried to get the gal to wait on the levy until I came back by car, but she insisted that if I were willing to try to get it the two miles back to the airport, she was too. I reluctantly agreed to let her stay in, especially since it was all open clear fields on the way. Luckily we made it with no further difficulty.

Problems were a little more serious another day, however. While on a return cross country with another duster pilot in the back seat, the engine suddenly burned a piston. Since we were at nearly 3,000′ and within gliding distance of the Davis Airport, I shut it off and dead-sticked in. It's hard to figure some problems. Unlike the dusters, the engine was in good low time shape, never abused, and was carefully maintained.

The make of engine didn't seem to be a dominant factor in failure statistics either. Since most of my approximate 10,000 hours in the air was in crop dusting with its tough test on engines, it's not surprising the P&W 985's and Continental fours lead the figures. Failures were not confined to them, however. For instance we had restored and flown a Culver

Helen prepares to fly our Culver PQ-8-A.

PQ-8-A for a while. It was powered by the Lycoming 0-290-B which was the military version of the 125 HP four. While returning to Clarksburg from a Tracy fly-in one lazy Sunday afternoon, I had barely gotten to cruising altitude when the oil pressure gauge began a slow, but steady, sag to zero. There was a very large dry stubble field of irregular shape, just ahead. Following my usual attitude of get 'em down fast while and where you can, I chose the near but short end of the field. It was an excellent choice as it turned out. A heavy piece of farm equipment had deeply rutted most of the field when it was wet, and I had landed in the only part of the field that the tricycle gear could have stood up to. The ruts had been undetectable from the air. A couple of phone calls soon had a supposedly like new replacement engine on the way from a pair of presumably reputable brothers who manufactured airplanes in Afton, Wyoming. Upon opening the crate, it was obvious that the engine was far from "like new" but at least it was complete and looked usable.

After installing the "new" engine with fresh oil and completing a preflight, a suitable take-off spot had to be found. The field I was in was far too rough, and short. With sloughs and crops eliminating any other nearby field, the only thing left was a dirt road on another good old levy. After a good run-up and mag check, I took off from the narrow levy and was soon at about 300' — and behind another lunched engine. Luckily, we were pointed right at a cropduster's strip by this time, and within gliding distance of it. After taking both engines home, I overhauled and assembled the best parts of each into one good engine and was able to finally get the plane home. The lazy Sunday outing had cost several hundred dollars, a couple weeks of sweat and a few extra heart beats.

Forced landings with round engines were not limited to P&W's and Jacobs. Helen and I headed for an outing to Santa Cruz in a 220 Continental powered PT-23 one day. We got all of about six miles and 1500' when the 220 announced its demise with much banging, belching, and smoke. It didn't worry me at all since there was no fire and a field I had already tested with a Stearman earlier in the month was right off our right wing. All incidents didn't end in powerless forced landings. One day I was test flying a friend's later model "Sesqui-wing" Waco Cabin that he had put together out of two derelicts. Right after take-off, I eased back the throttle — with no results. For the second time I had a throttle linkage come apart in an airplane. Luckily the throttle stuck near wide open this time. I eased over to the longer runway of Sacramento Muni Airport and by using the carburetor idle cut-off, I was able to blip my way down to a Nieuport-like landing.

It was the same runway of still another incident. I had given a basket case Davis DIW to a young friend of mine on the condition that he restore it to flying condition. He did a very creditable job, and installed a 145 Warner engine on the nose. I gladly accepted the offer of the first flight. It flew beautifully, but right after leaving the boundary of the field, the engine gasped and quit cold. Fortunately it returned to partial power long enough to make a 180 and land safely. It turned out to be just fuel contamination from insufficient cleaning of the old fuel tank. My young friend was Herb "Junior" Anderson who is now senior executive with the Pitts-Christen Eagle people.

As an active commercial pilot, requests for my services would occasionally come around that would be beyond my normal sphere of activity. Houston Walsh, a good friend from our sports car club, was an aerial photographer for a local Sacramento photography company. When their regular pilot moved on to better things, he asked me to temporarily fill in, flying their Howard DGA-15. Since there were many free hours in the dusting business, I agreed. The flying was easy for the most part. It was always done in clear beautiful weather over a good part of Northern California, an area that I was now quite familiar with. The only trying part of the flying was the actual photography runs themselves. It required holding headings and altitudes to fairly high standards.

Somehow the "temporary" stretched to quite a few months and eventually led to filling in a few times for a competing company flying their Cessna 180 and Super Cub. On occasion Houston would be unavailable to handle the camera and the boss would take over. He was a very methodical, by the book kind of guy. He would arrive at the airport with his arms loaded with maps, plotters, E6B computer etc. I totally ignored them. He would then ask me which way I planned on going. I knew the country by heart so I would just point and say "That way". I think it disappointed him when I got him right where he wanted to go. In spite of the Air Forces' many classes and refreshers on the E6B, I've never used or carried one in an airplane, ever. Doing this photography flying was strictly a moonlighting job and it began to conflict too much with the dusting business so I finally had to beg off.

People have always accused me of having a weird sense of humor and I suppose that's true. My buddy Scott Kemper had flown me over to another field in his old Lambert powered Dart to pick up my duster one day. The ancient engine threw a lot of oil and Scott kept a rag stuffed between the tubing and fabric of the cockpit to clean the windshield every so often. On the way back we were flying formation and, after one of his many wipes, a little bit of the rag poked through the fabric when he put it back.

I found it extremely amusing to see the wind grab the rag and slowly but surely start tearing a good sized strip of fabric back towards the tail of the airplane. It took Scott quite a while to figure out what all my laughing and pointing was about.

During my 25 years at the Clarksburg Airport, and the several years that preceded them, the ownership and management of the dusting business had changed several times.

The business was originally started as a partnership called Borges Brothers and Spezia. Vic Spezia soon decided to go on his own and established a dusting business a little south of Clarksburg in the Walnut Grove area. The parting was amicable and everyone continued to be good friends. The strain caused by the differences in temperament between the Borges brothers really began to take its toll after about ten years, however. Especially when Cannon made the snap decision to take over the Calexico airport. They agreed to split up their holdings with Joe taking over all of the farming and signing a five year non-competition agreement with Cannon, who then became sole owner of the dusting business.

By this time the dusting business had gone from a modest start of two or three airplanes to having as many as eighteen 450 Stearmans. However, the Clarksburg area was down to a one plane operation with the occasional help of a second airplane during the peak of the season. With Joe no longer in the business, and Cannonball moving to Calexico, some arrangement had to be made to handle Clarksburg, so Cannon suggested that Helen and I lease the business, which we did for a couple of years. By then the Calexico end of things was blossoming and Cannon was losing interest in our end so Helen and I wound up buying the airplane and the business.

Finally! My own business! I finally had it made — or did I? After several years of fighting insurance, telephone calls in the middle of the night, and dealing with some farmers who had trouble distinguishing the difference between "on Friday" and "by Friday", I had had enough and sold lock stock and barrel back to Joe, whose non-competition agreement had expired by this time.

Apparently, after a few more years of handling the business on his own, Joe decided to just stick to farming after all and he sold the whole works back to Spezia again. Spezia decided to jump back in with two great big feet and bought me a brand new Piper Pawnee duster to fly and hired a full time loader and right hand man. For two years I dusted in relative Cadillac heaven compared to the old Stearman. But after the two years Spezia could see that the returns didn't quite measure up to his investment and he wanted out. He offered to sell it back to Helen and me but we had already decided that I had used up enough of my nine lives. We not only declined the offer but decided to move on to something else. So ended 25 years of dusting.

I left the business with lots of memories, some good, some bad and some just plain amusing, like the time I was just standing in our driveway enjoying the beautiful day after a particularly tough morning's work. The fact that I was able to spend so many daytime hours at home compared to other workers was one of the benefits of the dusting business. Our youngest son Kip and a new neighbor boy, about three years old, rode up on their little tricycles and the new boy stopped right in front of me and looked me up and down without saying a word. He then slowly pedaled a full circle around me while he continued to look me up and down and stopped in front of me again. He looked me right in the eye and said: "MY Daddy works!" — and rode away!

FLYING RACERS AND OTHER EXPERIMENTALS

My appetite for the exotic and unusual had been deeply and permanently cultivated as a youth. Wherever we traveled after the war, rooting out and nosing around the local airports was always a high point of the trip.

Of course this frequently led to one of the occasional minor disagreements in our married life. It's well documented that men don't like to ask directions, but women do. Helen, being all woman, would want to stop and ask for directions to the airport while I would rather stop the car on the outskirts of town and get out and watch for an airplane entering a pattern. Being the driver I usually won. It hardly ever took long and, except for running into a dead end street once in a while, I almost always got us to the airport in due time.

In those days there was sure to be something of special interest on every airport. This was especially true of my two favorite fields in Los Angeles — Compton and Torrance Airports. Each was a hot bed of my kind of flying. So it was that driving up to Compton Airport one day brought the sight of a prewar racing plane to my unbelieving eyes. It was the Dave Elmendorf built Keith-Rider R-5 "Jackrabbit" and it was for sale! Wow! It was a spectacular moment! I just had to have it! It was a time when we were still adjusting to my new career in crop dusting and money was still a matter of some concern. No matter, I just had to seek out the owner to see if there wasn't some possible way of striking a deal. Amazingly he turned out to be more than willing to talk business. A person less star struck and impassioned with the prewar racers than I was might have picked up some warning vibrations from this and a couple of other indicators.

First, although the racer was still in perfect shape, it had obviously been sitting out on the flight line without attention for some time. Secondly, the price asked was a figure so low that it should have aroused some question. If I needed more reason to question the advisability of making a deal, I should have found it in the resulting negotiations with the owner. He was very anxious to move it, even to the point of accepting my ridiculous offer of $50.00 down (some of our food money, which hardly sat well with Helen!) and no schedule of payment on the balance of $700.00. No matter, I just had to have it.

Tickled pink with my new acquisition, I hurriedly filled it with enough gas to fly to Sacramento. I was very eager to get my prize home.

At that time, Compton Airport had a paved runway next to the ramp and hangars and three more parallel "runways" farther out that were hardly more than worn paths in the field.

Remembering the takeoff difficulties of some racers at Mines Field, I elected to take off on the dirt strip farthest from the flight line. I couldn't believe I was sitting in this vibrant prewar racer ready to take off — and it was mine!

I lined up angled a little to the right to compensate for the expected roaring torque of the powerful B-6-S Menasco. It had started and sounded so willing. The airplane had no brakes and with only a tail skid instead of a steerable tail wheel, I felt the takeoff needed all the help I could give it. Eagerly I shoved the throttle forward and braced myself for the fierce acceleration to come.

Nothing.

The combination of a very high pitched prop, a supercharger that needed quite a few RPM's to produce horse power, and tiny main gear wheels that had to struggle through a layer of soft dirt, caused all movement of the airplane to be held down to a fast walk. Since directional control had proved to be no problem, I decided to move back to the paved runway and try it again. Once more initial RPM and forward motion were quite poor, but at least it was accelerating some.

After a few hundred feet of slow acceleration, it finally reached a point where it all came together. Wow! It was like two mules kicking you in the back. It was off and airborne in nothing flat! What acceleration! But all too soon my exuberance was cooled by total silence from the engine. It had quit cold right over the wires. Fortunately it picked back up immediately but it would repeat this gasp periodically for the rest of the flight.

Then another distraction quickly appeared, the airplane began rolling radically from side to side and control inputs were having no effect. The salvaging factor was that the rolling back and forth was decreasing as the speed increased — only to be replaced by an ever increasing uncontrollable porpoising! Heeding the owner's repeated admonition not to fly it slow, since the lack of proper cooling at slow speed would soon seize the engine, I left the throttle in. The porpoising was progressing to the point that reading the instruments was becoming impossible. The last time I could read the airspeed, it was going by 220 MPH. The porpoise frequency was too rapid to compensate with stick control. It was so violent, that in spite of my being tightly belted in, the pitch over was throwing me upward to where the head high turtle deck was hitting me in the middle of my back. I was now over open country and if I had been wearing a 'chute, I would have simply unbelted and left the beast.

Hoping to find some way to stabilize things, I decided to pull the gear up and see if that might make a difference. Each time I hit bottom I was able to grab the ratchet handle and give the gear a couple of pumps. After a few butt pounding efforts, the gear was up but it had no effect on the porpoising. My next thought was that putting down some flap might change the center of pressure enough to make the difference. I managed to get what I thought was 20 to 30 degrees of flap down during the next few gyrations but there still was no change.

I might have found some humor in the situation under other circumstances. Here I was flying this bucking beast that was throwing my body half way out of the cockpit one second, only to reverse the force, and pound my butt on the bare metal seat the next. I was bounding in and out of the cockpit like an automatic jack-in-the-box. On each up swing, my goggles flew up over my forehead and then banged down on my chest when my butt hit the seat. It hardly presented the image a valiant fighter pilot would like to present!

At the time it was anything but funny. The violent action had affected my vision to the point that I could no longer see the airport, or much else, for that matter. The only thing left to try that I could think of was to see if I could pump the stick in rhythm with the porpoising. It had to be done by timing the rhythm since there was no way to react by feel. When I hit bottom the next time, I jammed the stick forward, and then back, then forward.

Viola! Finally I was flying straight and level and could see — thanks to about six inches of stick travel back and forth. I was over open fields and, in the interest of public safety, probably should have elected to make my crash there. However, I could now see the airport and decided help would be there if I was still in condition to use it after the crash. Just as I got to the airport my stick timing got out of sync and a horrible pitch over was followed by a spine compressing pull out — those wings were strong! Enough of that!

Elmendorf "Jackrabbit". Notice fuselage depth in comparison to me!

Somehow I was able to enter the landing pattern and with a decrease in speed the porposing also started to decrease — only to be replaced with that ever increasing roll from side to side. When I flared out over the end of the runway, the final roll to the left was so sharp that I thought the wing tip would dig in and cartwheel the airplane. Fortunately, the ground cushion stopped the rolling and I was able to make a very nice landing. However the roll out took me a little past the "runway" into the taller weeds where someone had dumped some trash and a lug box that poked a fist size hole in the right wing. Otherwise the landing had to be called a total success.

The owner hurried over and I told him there was no way I was buying that thing. "You can keep the 50 bucks for fixing the hole in the wing, but what you should really do is junk it." He replied that that was what he thought too, but when I came along and wanted to buy it he figured that was his best out. Only after it was clear that I would have nothing more to do with it, did he reveal more of the story. The pilot who first test flew it for him, landed it, threw his helmet on the ground and declared that he was not only through flying that, but was through flying anything! The owner said he often heard of people being white as a sheet, but that was the first time he ever saw anyone really white as a sheet. Fortunately, although the Jackrabbit was never a success before or after the war, it was one of the few prewar racers to survive. After passing through several hands without being flown again, it was donated to the E.A.A. Museum, restored, and placed on permanent exhibit. It can now be seen at Wittman Field, Oshkosh, Wisconsin.

The homebuilt (or amateur built) movement was an awakening giant in those early years. I would see my first post-war homebuilt airplanes at Compton Airport. Prototypes of the Smith Mini-plane, Stolp Starduster, and Stits Playboy were there. I was hooked. I had the good fortune of being allowed to fly all three of those some time later.

These were not the first homebuilts that I had seen, of course. While I was still learning to fly at Dycer a guy flew in a couple of times with a very tiny little biplane powered with a forty horse Continental. I also remember a used car dealer on Manchester Avenue having a tiny Heath Parasol along with another small homebuilt that were used as an attraction to draw in customers to his lot. It's amazing how men (and boys) have always been intrigued with tiny cars, airplanes, boats, bicycles or whatever! During this same period I was prowling around one of the San Fernando Valley airports when I saw a Knight Twister for sale. It was even cuter than the biplane that came into Dycer and I was really intrigued. I told my boss, Elwood, about it, maybe hoping he would buy it. He identified it from my description and said to never get near it, it was a killer design! I guess I never did listen to him much because many years later I flew Clyde Parson's 90 Franklin powered Twister. It was a real dog with that engine, but Clyde installed a 125 Lyc in it later on and that made a real airplane out of it. That era produced a number of other homebuilt airplanes of course and naturally some of the racers could be classed as "homebuilt" also, but certainly most of the racers were very professionally built.

Clyde Parson's "Knight Twister".

Paul Poberezny and the Experimental Aircraft Assn. can be thanked for the huge boom in homebuilt aircraft after the war. Goodyear racing had come and gone when I first yielded to the urge to build my own airplane.

Money was still tight which required much scrounging for parts and material. I designed a midwing racer to the Goodyear specifications hoping to get into racing if it ever came back.

I was able to buy a complete PT-26 for parts for $35.00. Although the wings had been cut in half for some unknown reason, it otherwise was just like new with only 50 hours on the airplane and engine! Would that those days would return! I immediately sold the engine for $50.00 and had many parts for my upcoming racer and other homebuilts for years to come. As design and building progressed, it was obvious that racing was not going to return anytime soon and many of the fine points of design for racing were given up for the convenience and comfort of sport flying. The required Continental 85 gave way to an available junkyard ground power Lycoming 0-290 I had purchased for $22.50. In spite of its heritage, it was a good engine, and with the addition of another mag and aircraft carburetor, it ran very reliably and made my "almost" racer perform very well indeed, much like a small fighter, in fact. The airplane gave many hours of great fun. On the way to fly-ins, Helen would fly the kids in our Piper Clipper we had at the time and when I would tire of formation flying at not much more than idle, I would speed up to cruise power and do lazy 8's across her nose with aileron rolls at the top of each loop. Most satisfying indeed.

"Macerschmitt"

Getting to that point was not without finding a couple of bugs in the first tests. Test flying an experimental airplane on its very first flight is not done without a lot of concern, not by me anyway. Nothing is for sure. The first bug showed up on the first short runway flights. The tail wheel I had rescued out of the airport trash bin gave up, resulting in almost running off the runway on roll out. So it was back to the shop for a better, if not new, tail wheel. I think something has been said about false economy.

I always pick the early morning hours for test flights. it gives cool air to an unproven cooling setup, and calm air to help check out unproven designs and controls. One highly unproven innovation in my design was the use of wing-tip ailerons. They caused the second bug to show up. I cycled the controls on the runway flights and all was beautiful except for the ailerons. They worked, but were quite sluggish. A light cross wind had come up by the time the tail wheel was replaced and we were back on the runway. I had gained considerable confidence in its directional control on the runway flights and had decided to go for it anyway. Right after lift off a gust of wind picked up a wing and neither a little aileron, nor a whole lot of aileron would bring it back down. The test flight was quickly terminated with a smooth one wheel landing and the wing tip about three inches off the ground. It was back to the drawing board.

Actually, I had anticipated the possible need for returning the ailerons to a conventional configuration so quick work with a skill saw on the wings and some work on the workbench soon produced a normal set of ailerons.

Then it was back to the airport. On lift off, I again had a wing lift slightly and remembering the previous experience, I fed in lots of opposite aileron fast and hard. She sharply rolled in response. I really had aileron control. It took a fast whapwhap in each direction before I settled down with the proper finger tip control inputs. With those short highly tapered wings, it had the fastest roll rate of anything I'd ever flown. From then on, the plane — now named "Macerschmitt" — was a huge success. It was flown by numerous pilots, including two ladies, one of them Helen. The wheels lifted off the runway for a total cost of $521.37.

The roll rate of this little plane led to confusion for the powers that be on more than one occasion. Rapid wing rocking was occasionally mistaken for aileron rolls. For instance we had become good friends with General Del Smith through the Sacramento Sports Car Club. He was the pilot of the California Governor's DC-3 at the time and we agreed that if we ever came across each other in the air we would engage in a little formation flying.

As a result when I saw an official looking DC-3 wandering back and forth near the airport one day, I quickly surmised that it could well be Del looking for me. I quickly jumped into my ship and flew along side the DC-3 — at a cautious, and legal distance — and rocked my wings several times. After getting no response whatever I wisely retreated. Later that day I got a call from a rather giggly Mel Ayres who was working the tower at the old Sacramento Municipal Airport that day. We were good friends through mutual interest in the E.A.A. and homebuilt aircraft. He said the DC-3 was an F.A.A. ship checking the radio range that was located half way between our two fields. When they saw me rocking my wings, they thought I was doing aileron rolls and called Mel in the tower and asked: "Do you know of any little airplane that might be up here in the airways doing acrobatics?" Since I was the only one around with a little airplane at the time, and without thinking much about it, Mel answered with a very accurate description of my plane. "Was it a little midwing?" "Yes." "Yellow and white with black stripes?" "Yes." Suddenly realizing that he just might be getting me in a whole bunch of trouble Mel responded: "Uh, gosh no, I have no idea who that might be!" There the understanding, and probably grinning, F.A.A boys let the matter drop.

Two more in-flight shots of Macerschmitt - a genuine joy to fly!

Mel told me of another amusing tower story. Aviation tends to be a small world and I had become acquainted with several airline pilots over the years. As a matter of fact one student I had soloed went on to become an airline pilot. For several years we lived under a long final for one of the runways at the old Sacramento Municipal Airport. When Mel was on duty one day, an airliner coming in reported his last position as being "over Harvey Mace's house". It was probably just mischievously done to see how confused the tower operator might become but the airline pilot was probably surprised when Mel came right back with: "Roger Flight 321 two miles out, you're clear to land." I never did find out who the pilot was.

We had been bitten heavily by the homebuilt movement and I was one of the early members of the Experimental Aircraft Association, member #2714.

When word that I was building an airplane spread among the aviation community, there was quite a string of visitors to the shop. Most were in the "Gee, that's nice," group, but there were many that fell into the " Why don't you ________ ?" category. "Why don't you put two wings on it?" from the biplane enthusiasts. "Why don't you give the wings a high aspect ratio?" from the sail plane group. Or "Why don't you do that this way?" from the know it all group.

Scott Kemper was a buddy since my BCRA racing days and was an enthusiastic supporter and fellow pilot. One of his coworkers was an ex B-29 pilot named George Pereira. He once had to bail out of a burning B-29 from combat damage on the down wind leg of his home base.

George was not only interested in my project but was also interested in building something of his own. I helped him with his first project and he proved to be a fast learner. His wife, Peggy, agreed to allow him to start building a single place seaplane with the provision so that I do the initial test flight. I readily agreed even though I had never even had a ride in a seaplane, and besides, most first time builders take years to finish their projects, if they ever do. At the time I thought the question of who would fly it was strictly academic.

I hadn't counted on George's ability and resourcefulness. He had the project done in short order and I soon found myself sitting in the cockpit in the Sacramento river's deep water channel. I made the first short flight down the channel and back and turned over further flights to George. Scott was also there, he had helped some on the project and we all celebrated George's success with hand shakes all around.

George Pereira's "Osprey I" as test flown by author.

The Navy showed great interest in the airplane because of its possible use in the Viet Nam war. They bought it and it now sits in a Navy museum.

George had a much better head for making money than I do. His next project was a two place amphibian of very nice lines that made quite a splash (sorry) in the homebuilt field. George sold plans to this airplane and made a pretty good piece of change.

A swap made entirely over the phone was the cause of one of our many personal plane changes. I traded our Piper Supercruiser for a C-90 powered Stits Playboy. We agreed to meet at an airport approximately half way between our home fields and managed to arrive within minutes of each other. We signed the necessary papers without ever getting closer than 100 yards of each other's planes. We then went our separate ways quite satisfied with our new toys.

I had never given up on aerobatics and hoped the Playboy would be the answer. It was a good flying airplane but the only maneuver it was any good at, other than things like Lazy 8's, was a snaproll. It would do single or double snaprolls really well and could be stopped right on the point.

While doing aerobatics one day, I finished with a double snap and found things very wobbly on recovery. A quick look at the wings revealed the fact that the twelve inch aluminum wing strut fairing on one wing had popped loose and laid against the strut 90° to the wind. It made a perfect spoiler. No amount of rocking around would shake it loose. It took considerable aileron to hold the airplane level and I feared for the prospects on landing. With no parachute, there was nothing to do but try, however. While slowing down during the landing flare, it took more and more aileron travel until there was no travel left. The wing just kept going down. It wound up being a race to get some kind of a landing with it before the wing got low enough to hit. Luckily I was able to make a good one wheel landing with the wing tip just skimming the grass. Needless to say, the fairings were replaced with smaller fool-proof setups before the next flight.

Helen had taught herself aerobatics in the Playboy and the only other minor incident happened when the cushions she had to use put her shoulder too close to the canopy release. While doing a snaproll, she bumped the release with her shoulder and away went the canopy.

Somewhere along the way we became the owners of a damaged, engineless, Stits Flutterbug. I rebuilt it and mounted one of my now proven ground power 0-290's on it. It wasn't good for much, but it was fun to fly anyway. We enjoyed showing off its short field takeoff ability. We managed to fly both our airplanes to several fly-ins while we had them. At the Merced Fly-in, I won the short field take off contest at 105', on a near 110° calm day in that Flutterbug.

On one of my airport prowlings at Compton (which was also the source of our PQ-8-A) a gaze through a musty hangar window revealed the fuselage bones of what must have been a Grieve Class racer. It wasn't familiar to me, but it just had to be one. Investigation revealed it to be the remains of the late Art Chester's "Goon". Although it was in the care of people we would soon call friends, it still belonged to Trudy Chester and could be bought.

The people who were storing it for Trudy turned out to be Al and Joanne Trefethen, both avid flyers and builders. We became good friends for many years to come. Al trailered many strange things up to Sacramento over the years, including our PQ-8-A.

The "Goon" wing was kept at the Trefethen's ranch at the foot of Palos Verdes. When all of the parts were gathered together, it proved to be very complete and undamaged. The engine was missing, having been donated to the Northrop Flying Wing Project during the war. It was lost when one of the small flying wings went in. Even with the engine and some hydraulic gear retraction parts missing, it was still quite restorable. We bought it and trailered it to Clarksburg. We were crossing Art Chester's path once more.

Top: Photo of Art Chester's "Goon" taken during the filming of a pre-war movie. (John Caler photo)
Above: Nice profile of "Goon" with Art preparing for flight. (EAA Archives)

I'm sorry to say that with no shop to work in, and precious little money, my efforts didn't equal that of the original builders. But I did save the airplane, did restore it, and did have it ready to fly again.

I was soon on the long runway of Sacramento Municipal Airport, the same old runway of several of my other escapades — and wondering how I kept getting into them. I roared off for the first of several runway flights. I needn't have been concerned about flying it. It was a beautiful handling bird. Takeoffs were extremely straightforward and landings were grease jobs.

Of course in threepoint you were blind as a bat. All of that wide 6000' runway would disappear behind the nose. The many hours I had in P-51s, P-47s, and 450 Stearmans made that quite acceptable, however.

Everything was ready to go for the first full flight but I decided to wait for the cool, calm air of the next morning. But the next morning brought fog — along with the next day, and the next. For thirty days and nights we had fog. Well, we would just have to wait.

Inquiries everywhere turned up only three Menasco C6-S engines, all in museums and unavailable. I bought the airplane to fly, so even though it hurt, I changed the engine mount and installed an 0-435 Lycoming and variable position prop. I was careful to make the engine mount change so that it could be restored to original if ever an engine turned up. I made an entirely new cowling so as not to damage the original. This plane was masterfully built, obviously by professionals all the way.

Art Chester was considerably smaller than I was. Even without cushions I couldn't close the canopy. Remembering the flight in the Jackrabbit, I didn't care to fly the Goon away from the runway without a 'chute. A seat pack was out of the question and a backpack not as comfortable as a chest pack, so the chestpack was what I decided to use. It meant crooking my arm in an unnatural manner to get at the stick but it seemed to be O.K. sitting there on the ground.

As soon as I got airborne, I knew I had made a mistake. On lift off I was all over the sky, relatively speaking. The unnatural arm position destroyed all the delicate feel for the stick that this thoroughbred demanded. I quickly banged it back on the ground and went back for a back pack. It was uncomfortable but far more suitable.

The first full flight then went without a hitch. It was indeed a beautiful flying machine. It's only drawback was a very poor rate of climb with that lower HP engine, which was less than half that of the C-6-S. In fact with gear down, the rate of climb was almost non-existent. That necessitated getting the gear up immediately. Having to pump the gear with one hand, while holding the responsive stick with the other, made for some interesting gyrations right after lift off.

I gained immediate confidence in the Goon, however, and made all subsequent flights without the 'chute.

A big E.A.A. fly-in was coming up soon at the Riverside Airport. It meant a long cross country flight for the relatively untried Goon, but I was eager to have the Los Angeles gang see it. I took off for the fly-in with no more than two hours on the airplane. Although Los Angeles was nearly 400 miles away, there were many airports right on course. With the speed of the Goon, I would never be more than ten minutes from an airport and the rest of the country was good for bellying in — except over the Ridge Route. There, stretching for about 20 minutes, would be inhospitable ground indeed, and there, right in the middle of it, is where the Lycoming decided to gasp. it was the only time it did in several hours of flying. With my heart racing the RPM it did keep running, however. I flew on to Torrance to join up with Trefethens and the rest of the gang. We left for Riverside in a variety of experimentals that included the Goon, Mini-plane, Poopsiedoll, Sizzler, and Trefethen's Sportaire.

When we arrived at Fla-Bob, I was somewhat taken aback by the sight of the runway. Somehow, it sure didn't fit the description that I'd heard. It was long, and paved all right. But the pavement was only about 20′ wide and covered only half the runway. Well, I'd come all this way so I had to go for it. With a carrier type approach and a last minute slip, I was able to keep the runway in sight until it was just barely time to straighten out and flare for the landing. At that point, the runway completely disappeared behind the nose of the Goon and I had to use a barbwire fence parallel to the runway as reference for the balance of the landing and roll out. It worked, but I wasn't about to tempt fate twice. The crowd would just have to wait to see it fly again when we left for Torrance. At that time I gave them a couple of good buzz jobs and hoped that would do.

Eventually I was heading back to Sacramento with Al flying formation in Rudy Hazuka's AT-6 taking pictures. As we neared what used to be good old Mines Field, now a many runwayed bustling LAX, the Goon's engine quit cold! By quickly pulling carb heat on and jazzing the throttle the engine thankfully started running again. I'll never know what the cause, or the cure was for sure but by then I had lost my confidence and decided to land at LAX, radio or no radio. Besides, I knew a fellow in the Los Angeles area who would kill for the chance to buy the Goon. I made a high, close approach, making sure there were no airliners to interfere with — amazingly for LAX, there weren't any in sight — and landed very near the spot where I had gotten my solo license years before. I parked in about the same place that we had kept the Rearwin and Cub. Everything was completely changed, of course. I made a quick call to the tower to throw water on the heat I must surely have generated up there.

Two photos of "Speed" Gardenhire's air utility cargo A/C. Photo at left shows the cargo pod in front of A/C, at right is author, airborne.

"What? uh—oh—uh yeah, oh—uh that's O.K.—uh —— no problem." It was obvious that they had seen no more than the C.A.A. inspector had when he gave me my solo license years ago.

I hated to sell the Goon, but practicalities dictated that I should. Anyway I had flown it quite a bit, and it was time to move on to something else. My affinity for the weird and the wonderful, and my willingness to test them, sometimes led to interesting events indeed.

A non-pilot mechanic friend named "Speed" Gardenhire had conceived and designed a twin boom monoplane with a detachable pod that could be adapted to many uses. He built a flying model of the idea powered with twin 65 HP Continentals. Without the pod, you might say it was a stand off model of the P-38 — if you stood way off. I made the first, and subsequent, test flights off an old Air Force Auxiliary field south of Sacramento. After I felt comfortable with it, I flew it to Clarksburg where it would be easier to work with. It flew fine except for a slight buffet in the horizontal. Speed repositioned the horizontal to cure the problem, and it was soon ready to try again.

Right after lift off, I noticed the need for more and more forward stick. It wasn't until after the short runway had fallen behind that the mental alarm bells really started ringing. The airplane continued its increasing climb until I ran out of forward stick travel. The only option left was to come off the power. I was just above the tree tops when I found that slowing down to just above a stall allowed me to maintain level flight. This with the stick against the forward stop!

My first reaction was to put it down in a field immediately, but since it seemed stabilized, the gamble to get it back to the airport didn't seem unreasonable. After a very gentle duster type turn to reverse direction, I was able to throttle in for a very nice fourpoint landing (two mains and two tail wheels) with the stick never leaving the forward stop. Some readjustment of the horizontal was certainly in order.

I am quite used to flying airplanes that I have had no previous experience in, having been called upon to do so many times. Of the over 75 distinctly different aircraft types that I have flown, only eight were flown after a dual check out and a similar number after a verbal coaching. For all of the rest, I just had to get in and fly them. This includes airplanes like the P-47, Johnson Rocket, Stagger Wing Beech, and of course, all ten of the airplanes that I made the first test flights in. In addition to the ten totally new aircraft, I made the first flights on at least a dozen heavily modified airplanes.

My reputation for flying anything seemed to be spreading. One day I got a phone call from someone in town I had never heard of. He called to ask if I would please test fly the airplane he had built. "Well, I suppose I would, what is it?" As he answered my questions of configuration, empty weight, wing area, power, and propeller pitch, the picture formed was an abomination. It obviously wouldn't fly, and I said so. "What do you mean it won't fly!" he must have bounced off the ceiling. He also sounded hurt, so I agreed to look at it anyway.

When I reached the field, it didn't look too bad from a couple hundred yards away. Maybe I had been too hard on him. As I got closer though, each new step brought into view some new horror — it was far worse than I had expected. The little 40 HP engine was hanging from a solid piece of 3/4" exterior plywood backed up by another 3/4" piece of solid plywood, backed up by a third piece of 3/4" plywood which, mercifully, did have a couple of lightening holes in it. Those holes were then covered with a thin piece of plywood. The seat and seat back were made of — you've probably guessed — solid 3/4" plywood. The seat had ended up a little too low for the builder's satisfaction, but that was no problem for him. He took a couple of 2X4's and laid them on the seat and covered them by still another solid piece of 3/4" plywood, and that quickly solved that.

The tail surfaces were also cut out of 3/4" plywood with lightening holes, then covered with thin plywood or fabric. The main keel, or backbone, of the fuselage was originally designed to be made like a box spar, but since the builder weighed well over 200 pounds his answer for more strength was to band saw the keel out of a solid beam, the original dimensions of which I was too horrified to contemplate. The banana shape of it put the grain of the wood running off the beam right where the gear legs and wing struts attached. Poor engineering indeed. Some of the tubes used (push-pulls etc.) were made from thin wall conduit, mashed with a hammer on the end to put a bolt through. The wings were covered, so what lay under the fabric was mercifully hidden from view. The size of the fin and rudder were much too small for a light plane, they would barely have been adequate for a 200 MPH racer.

The brakes were actuated together by a hand lever. This, in company with a full swivel tail wheel and small rudder, made even taxiing completely out of the question. The spring steel gear legs were so under strength that a light touch on the wing tip made the plane rock like it had been hit by a strong wind. I heard that when the F.A.A. came out to look at it for licensing, he turned to the builder and suggested that he not give up his day job. Of course the inspector wouldn't license it.

The airport was pretty deserted so I agreed to run it down the runway if he would make the tail wheel steerable. After all, how much trouble can you get into with something that won't go very fast, much less fly? He soon called to say it was ready.

I taxied out, rocking to and fro, and poured on what coal there was. At least the gear location was right and I was able to get the tail up. After reaching "full" speed, I was curious to see if I could bounce it off the runway even a little bit. No way, the tail wheel would bang down with a thump and that was it. When I felt I had thoroughly satisfied the builder that it wouldn't fly, I was ready to throttle down. Suddenly she cocked hard to the right as if the right brake had locked and in a flash she was on her back. It was not a hard flip, but of course many things broke, including the "over-strong" back bone. Mercifully the owner conceded defeat.

After I crawled out, the first one to reach me was the builder's buddy. He rushed up and shook my hand. "Thank you! thank you for doing that! He would have ended up killing himself in that thing!" He thought that I had done it on purpose! There was other scuttle butt that had me wrecking it on purpose. But I didn't, honest!

Eventually the Goodyear type racer, and racing, made a comeback and my interest was rekindled. Even more so when Al Trefethen had me build him a racer using an old wing he had. He enjoyed racing a couple of seasons with it and it continued to race quite a few more seasons with a new owner.

A month after returning from the 1969 Reno Races, Helen and I were having coffee, hashing over the event, when I casually suggested that it would be fun to build a racer for ourselves. When she agreed, I immediately got out the drafting equipment one more time and set to work. I wanted to be a contestant in the Reno '70 races. It was a lot of work, both on the drawing board, and in the shop. About two weeks before Reno Races '70, I had it ready to test fly. We now had a brand new racer named "Shark," after all those years we were going racing. Almost.

First we would have to get by the preflight F.A.A. inspection, get over the first flight hurdle, debug the glitches that would surely show up, fly off the required ten hours, get by the second F.A.A. inspection and flight demonstration, get it up to Reno early for the P.R.P.A. technical inspection, and fly a pilot proficiency demonstration. If all this went well, we would then be allowed to try and qualify for a spot on the program — there were always more entries than the program could handle. Such optimism!

Fortunately, the initial F.A.A. inspection went well. The inspector barely glanced at it on the way to the kitchen table to fill out the paper work, and the only glitch in the plane showed up on the first runway runs in the form of some unwanted toe-in of the wheels. This was easily cured and we were soon on the end of a runway, one more time, ready to take off in a newly constructed airplane that had never before taken to the blue. It was mixed emotion time — I was glad I was doing it, but sort of wishing I didn't have to!

The flight couldn't have gone better, it flew like a dream and seemed quite fast. Within ten minutes I was doing aileron rolls across the sky. Time was running close, I would have to get the F.A.A. down right away. Good old Parker Pen helped get the ten hours off very quickly. Finishing the demo flight for the F.A.A. with a tight turn around the wind sock and a fairly low altitude roll convinced Rysaavy it was ready. We were off to Reno after eleven months of work.

I couldn't have had a more supportive crew than Helen and our sons, Lan and Kip. Though quite young at the time, Kip was a most enthusiastic helper at the races; and it was Lan's help with the paint and putting the engine together that really helped make the deadline. It was a trouble free engine.

I had seen the disappointment of other first time race plane builders who had set their sights much too high so I only hoped for a qualifying speed of 200 MPH on the closed course. This was no where near as fast as the fastest racers, but it was a feat accomplished only once before for a first time racer. That was John and Joanne (ex-Trefethen) Alford's "Pogo". As for the races, placing high in the main event was out of the question. Unfortunately there are people who cheat and we were running an absolutely stock engine. I did have hopes of winning the semi-main called the Silver Race.

The P.R.P.A. tech inspection and demo flights went well, and we were ready for the

The Mace R1 being raced by Bud Pedigo (#98) and eventually losing to the Mace R2 being flown by Bill Skliar (#41) in the Mojave Races of 1975. Bill Skliar photo.

The Mace R2 "Shark".

qualifying runs — and the Shark still didn't have over an hour and a half flying time total. On the qualifying run with full throttle and flying as smoothly as I possibly could, we rounded the pylons to the checkered flag. A quick check with the timers after landing showed we had done 198.8 MPH. Not quite my hoped for 200, but close enough. We were on the program and comfortably so.

Takeoff for the Silver Race was memorable. The racer on my left was half a plane length ahead at lift off and drifting into my lane. The racer on my right was far enough behind that I couldn't see him out of the corner of my eye. The one on the left kept squeezing closer. I love formation flying so I held my line, thinking how surprised he would be when he glanced over. He never glanced, and when my wing tip was dangerously close to touching him, I knew I would have to give way or hit him. I glanced over my right shoulder to be sure I had room and moved over. That racer was comfortably clear but the eyes of the airline pilot who was flying it were as big as saucers.

It was really exhilarating. After living a half hundred years, I was really racing at the National Air Races.

I didn't win the Silver Race, but it was close, I took second behind a main event racer who had dropped down to the Silver Race by virtue of engine trouble on qualifying. I was satisfied, we hadn't set the world on fire but through my eyes it was quite a bit brighter.

The excursion into racing was also planned as our farewell to commercial flying. After 25 years and many forced landings in the crop dusting business, I felt I had pushed my luck about as far as I had dared. In the early days, a routine forced landing was looked upon as a conversation piece; while the really close calls might be cause for jumpiness no longer than a few minutes. Age tends to breed conservativeness and a stronger appreciation of the years that might be remaining. After a really close call, it was getting to where it might be two or three days before I would quit jumping at every little rattle or gasp of the duster.

We sold the "Shark" and our Luscombe 8-E and bought a 40' commercial fishing boat. The "Shark" eventually ended up in the hands of Bill Skliar, an ex-test-pilot on the SR-71 and big wheel at Edwards AFB before his retirement from the Air Force. Bill was a heck of a nice guy and raced the "Shark" at Reno for a number of years. Eventually high aspect ratio wings were the in thing to have at the races and Bill and his crew built a set for the "Shark". Most unfortunately the wings failed on a test flight and Bill was killed. The remains of "Shark" were buried right on the race course. The original outer wing panels are now in a museum on the field.

For the first time in many years, we were completely out of a means of flying and perhaps for good. But, not to worry. We tried, but couldn't put up with that for long and soon went in partnership with Al Trefethen on a good old Taylor craft. Al was divorced, retired, and had also moved to Fort Bragg by this time.

In the meantime we had also become friends with a fellow E.A.A.er down the street who was building a Volkswagen powered monoplane. Due to his lack of flying experience, he asked if I would do the test flights when it was finished. Never able to resist an adventure, I agreed. In a very short time, I once more found myself on the end of a runway ready to take off in an untried airplane. The runway was plenty long, but the field was completely surrounded by forest, a forest that had already claimed a number of hapless spam cans over the years.

Takeoff in the Volksplane was uneventful and it seemed to be flying in good order. Experience dictated a flight path within convenient gliding distance of the runway, however. The wisdom of this precaution was quickly underscored when at 1500′ the engine coughed, sputtered, and reduced its power output to a rough idle.

When would these forced landings ever end? With a mile long runway within gliding distance this one would be routine, however. As it turned out there really was a bug in the engine. A little bugger was blocking the main jet. After cleaning out the carb, further test flights were uneventful. Eventually tiring of the Volksplane the builder sold it and began construction of a KR-2. In the meantime I had tired of the struggle for bucks in the fishing business and decided to put my talents at building airplanes to work.

George Pereira's amphibian had proved to be a large success and a number were being built all over the world. There were two prospective customers who just didn't have the time to build and George steered them to me. An agreement was reached to build the two amphibians for the two fellows: I was to finish one through flight test and the other up to cover and finish work. As with building any airplane, the task was arduous and time consuming.

Eventually, however, I was once more ready to first flight an untried homebuilt. This time it was an amphibian. It would be my second flight off water. After an adjustment to its step angle, it performed quite well. My total seaplane time was now around one hour, all solo. After several takeoffs and landings it was turned over to George who proceeded to give quite a demonstration including tight step turns.

Problems in flying can come at unexpected times, from unexpected sources. It was at about this point that we found a derelict Ercoupe with flat tires and in need of a paint job sitting at a small airport. Close inspection showed it to be basically sound and worth saving, however; and it was for sale at a very reasonable price. Although it had been in long time storage, the owner assured me that it had flown into the field it sat in. Nothing in the inspection turned up anything to indicate that it wasn't flyable, so a deal was made. A ferry permit was easily obtained to make a five minute flight to a friend's airport where I would have better facilities to do the needed annual and relicense before making the hour cross country for home.

Run-up showed the 85 Continental to be running strong with only a couple drops of oil on the windshield. That was not all that unusual since it had sat so long . Although the tires were flat, they were new, and some air put them back in business. In due time we were ready to go.

I had flown Ercoupes before and was impressed with their good performance for the horsepower. They are better airplanes than most people give them credit for.

As I reached the end of the runway, I noticed some workers nearby. Always the ham, I decided to show them an Ercoupe's maximum takeoff performance. After a minimum takeoff roll, I horsed it into the air for a maximum climb-out and all hell broke loose. That Ercoupe was uncontrollably all over the sky. My first thought was to get it down while there was still runway left. But before I could do that I had to get it into some semblance of level flight.

It was a struggle, but with reduced power and gentle control manipulations, she finally settled down. There was still lots of runway left but since it seemed somewhat manageable straight and level with mild control inputs, I decided a landing at my destination would be no worse than here. As the airport fell behind, the one or two drops of oil on the windshield began to increase — she would definitely need a new nose seal. By the time the short flight was nearly over, the windshield was too covered with oil to see through. This was not really a problem with the side windows down, however.

Left: Skyhopper we owned for awhile. Right: The wing of my last home-built under construction.

A very gentle, slow pattern brought the Ercoupe onto a drag-in final. A little additional power to clear the fence and I would have it made. Drat! Even with the carb heat on she was icing up. It was going to be close. As if that wasn't enough, some low altitude turbulence necessitated more forceful control inputs. This brought back some more of the wild gyrations that had occurred on takeoff. What in the world was going on? One wing was almost touching the ground when some diligent two handed work on the control wheel finally got the wheels on the runway. But steering was wild during the rollout and it was very difficult to keep straight.

At last I got it parked and no harm was done. I had parked it in a turn by the hangar and while walking around it, kicking the dirt and wondering what in the world could make a docile old Ercoupe such a handful, I happened to notice that while the nose wheel pointed right and the right aileron was up as it should be, the left rudder was pointing left as it should not be. Someone had hooked the rudders up backwards! Also a close look at the carb heat butterfly showed it missed closing by an inch. No wonder it wanted to ice up on the approach.

A new seal failed to stop the oil leak, and while I was trying to figure the why of that, a disgusted push on the shaft brought some movement, movement that shouldn't have been there! Later disassembly would reveal a .010 undersized shaft in a standard size bearing. No wonder the seal wouldn't hold. I made a quick trip home and removed the C-85 out of our homebuilt Skyhopper and installed it in the Ercoupe and soon had things in order for relicensing. Eventually we got everything home.

The Ercoupe is such a nice easy airplane to fly that our youngest son Kip took to it like a duck to water. He could have soloed it in thirty minutes time had he been properly licensed.

The saga of the substandard work on the Ercoupe's engine was not yet over, however. When I disassembled the engine for new bearings and rings, I removed the oil tank and sat it upside down on a clean board to let the last of the old oil drain out. I figured that by reassembly time it would be clean enough to just bolt back on. That was my mistake as time would tell.

Since both engines were about the same age and condition, I elected to leave the Skyhopper's in the Ercoupe and installed the Ercoupe's engine on the Skyhopper. We then sold the Skyhopper to Scott Kemper and during the delivery trip over the mountains, the oil pres-

"Scorchy" and I out for a little fun.

sure gauge wobbled and said all was not right. Since I had just passed an airport I turned back and landed there immediately. One more forced landing — when would they ever end? At least on this one I still had power and a handy airport. Disassembling the engine would reveal a handi-wipe left in the oil tank by whatever inept mechanics had been working on the Ercoupe. Of course, that includes me as I have to shoulder some of the blame for not flushing the oil tank when I worked on it in the first place. Live and learn.

Our oldest son, Lan, came across the discouraged owner of one of Jim Bede's BD-5 kits. It was in unopened boxes and could be had at a considerable discount. Lan was interested but unable to swing it alone. I thought it would make an interesting father and son project so I agreed to go halvers on the deal. I wasn't interested in it just for myself but, naturally, things don't always go as planned. A marriage by Lan and a chance for him to buy some property found old Dad as sole owner of the BD-5 kit anyway.

Jim Bede found himself in the throes of disaster soon after this and folded. I, along with hundreds of other BD-5 kit owners, found myself with a shaft-driven, pusher-propelled, mid-engine hot rod with no engine or drive train. Having grown up in the Los Angeles hot rod element where putting strange engines in strange places was commonplace — I had done my share — it didn't seem unreasonable to seek out an alternative solution for getting the BD-5 into the air. I elected to make the tri-geared, mid-engine, pusher into a tractor-engined tail-dragger. There were a lot of skeptics over that idea but in due time, the little low wing all metal hot rod was ready to go. Ernie's KR-2 was also ready to test about this time. Both airplanes had Revmaster 2100-D engines for power.

After six runway flights that included four grease job landings — number's three and five were real bummers — it was time to get it up and away. But, I couldn't get it up and away completely scott free. The engine stumbled right at the edge of the trees, but did keep on

The clean, functional lines of this tiny A/C are highlighted in this view of "Scorchy".

running. Another heart thumper, but at least the rest of the flight went beautifully. A needle adjustment on the carb cured the stumble and the airplane proved to be a joy to fly.

I was soon enjoying the exhilaration of smooth lazy eights and barrel rolls. For a 60 HP 450 pound airplane the flights were amazingly reminiscent of the P-51, especially one day while climbing up through multiple layers of broken clouds over the coast. It brought back the feeling of heading out over the English Channel as I had done so many times nearly 40 years earlier. Only this time there would be no crossing and there would be no flak. Even being in my senior years couldn't keep me from doing a 210 MPH buzz job and victory roll when I returned to the airport.

Ernie Stumpf felt unprepared to test his KR-2 so I once again had the finger put on me. For the second time in three days I found myself sitting in a totally new untried airplane at the end of this tree rimmed runway — and still wondering how I keep getting into these situations. The test flight went without a hitch, however, and it proved to be a very nice flying airplane, much like the Culver Cadet that we used to own.

We kept "Scorchy" quite a while and took it back to the big E.A.A. gatherings at Oshkosh two years in a row. I flew it during the fly-by parts of the show and managed to raise a few eyebrows. That was a lot of fun. In spite of the enjoyment that Scorchy brought, my tendency to lose interest in things with time — or my continual desire for something new, or both — caused me to wind up trading it for a Piper Vagabond.

Eventually my penchant for swapping things around found me in possession of a whole gang of parts from an unfinished homebuilt project that our friend from Canada had started and lost interest in. It consisted of a modified Jodel fuselage and a 0-time since major C-90 engine and metal prop. We had a Cessna 150 at this time, which is a really nice airplane but quite a ground lover and not noted for a rocket climb. I felt that I could build a better performing two-place putt-around.

Using the Jodel shell as a start, I built a three piece wing with minimum dihedral and no twist in order to get the maximum lift possible. If the stall turned out to be too severe, a couple of stall strips near the wing root could take care of that. After a year or so of work, it was reluctant test pilot time again. I had little concern about the engine, it and all the accessories were "0" time since overhaul and were running great in spite of having been in storage for some 20 years. Consequently I decided to forego my usual demand for a long runway for a test flight and decided to fly it out of our 1700′ local strip. I knew the takeoff wouldn't use much runway as my first fast taxi effort got me a quick surprise lift off — followed by a very quick landing since the fog was down on the trees! This was followed by several intentional runway flights so I knew I had my short field airplane.

The first full flight confirmed its short takeoff and climbing ability and was uneventful until it was time to throttle back for landing. It then became quite clear that something was amiss in the idle system, the engine showed every intention of quitting in that mode. A long drag-in approach and letting the engine die after clearing the last of the trees made the landing come out all right, however. An offending piece of some kind of solid material removed from the idle jet soon put everything right. There have been many successful flights made since.

Among the pieces needed to construct the plane were a couple from the last remaining bits of the PT-26 purchased for parts to build the Macerschmitt over 30 years before. About all that remains of the PT now are a couple of very short bits of tubing and a push rod or two. Whether they ever find their way into future projects remains to be seen. There are at least six aircraft flying with bits of that PT-26 in their structure.

Flying the homebuilts has given us a world of pleasure and a range of experience that would be hard to duplicate. They furnished fond memories that will be with us until the final closing of the hangar doors. Regretfully, we sold our last airplane, our homebuilt, a few months ago. We had flown it for over five years, which is a record length of time for us to own an airplane. We are currently without something to fly, but hopefully that won't last long.

The war took the steam out of the desire to fly for many WWII pilots. It is certainly understandable, especially since there were a lot of them who weren't all that fired up about flying in the first place. At least not in the sense of being star struck over aviation as I was.

One of the few charter members of the 362nd Fighter Squadron that did keep up with his flying was Lynn "Junior" Drollinger. He was not really a junior but got tagged with that nickname because of his slight boyish build and young age. He still answered to Junior to those of us that are left out of the old squadron, right up until the tragedy of his fatal heart attack. He not only kept up with his flying but was also into the E.A.A. homebuilt movement and had built two gorgeous two place aircraft and was nearing the completion of a third at the time of his death. He will be missed.

FLYING SAILPLANES AND OTHER STRANGE THINGS

The silent, seemingly effortless, soaring of sailplanes had not escaped my attention over the years. I had watched them with envy many times. I had even towed one up with the Stearman duster on one occasion. But there's just not enough hours in a day to do everything one would like to do.

One of the amphibians I built was for a fellow who operated an airport and specialized in sailplanes. He talked me into taking the time to fly one. He put me into his two place Schweizer trainer and away we went. After 15 minutes and two landings, he got out and said, "Go for it!"

It was an absolute ball! After many, many engine failures in power planes, where the only option was down, it was quite a thrill to cut loose from the tow plane and see "up" on the rate of climb — with no engine at all!

After that second flight in the 2-33, he talked me into the single place 1-26 and that was even more of a blast. I had some very enjoyable hours soaring around the foothills. I looked over the used sailplane market with a fond eye for quite a while after that, but the area we live in is just not sailplane compatible.

One type of flying I would not recommend attempting without some coaching and practice off "bunny hill" sand dunes is hang gliding. My oldest son Lan, and his buddy, had taught themselves to hang glide off a small local sand dune and had graduated up to the high hills and cliffs. I don't think anyone else within 100 miles was flying them at the time.

I watched them with much interest but avoided joining them because I was already up to my neck in other hobbies. After watching their big grins and enthusiasm for two years it finally became too much for me to ignore. The local sand dunes the boys had used were still available so we took the hang glider there. About the best that anyone could squeeze out of this "hill" was a flight of about 40′ at maybe 10′ high. The sand and size of the hill virtually precluded anyone getting hurt so it was an excellent spot to get started.

My first two efforts clearly pointed out the folly of anyone jumping off of anything higher without some practice and coaching. Particularly a habit prone pilot. It took 13 jumps off the Bunny Hill before I felt ready for the big one. I was already in my late 50's and bones do get brittle.

The hill we moved to was about 300′ high but when I looked down from the top, it looked more like 1,000′ to the bottom. I took a death grip on the hang glider, bit the bullet and ran off the hill. With the sail ruffling from the excessively safe speed I was holding, and not daring to look at anything but my landing target, I made a "sled" flight down to a passable landing. It was really a gas.

I hurried back to the top for more. By the third flight, I was making gentle turns. This was much closer to flying like birds than anything I had done before. I was soon eager for a bigger cliff. The only one available within a reasonable distance was at Usal. It was about 700′ high and made for fabulous flying. Generally, conditions were only good for six or seven minute runs. But a day did finally arrive that provided the right conditions for true soaring. Instead of the usual 20′-30′ drop at leapoff, an instant surge carried us up. We were soaring. At one point I turned into the wind, put my shadow on a rock and held it there, never moving. Flying like the hawks would never get closer than this.

Getting to this point was not completely incident free. On "no lift" days a vigorous run to the edge of the cliff is necessary to insure having adequate air speed for the takeoff. On one of these occasions, I tripped on something right near the edge of the cliff and went off into space in a very disorganized attitude. I managed to save the day but only after bouncing my butt off a ledge about 40′ below. It was painful but I managed to be grinning by the time I landed.

On another occasion the wind direction made flying that day a bit questionable. I volunteered to be the "wind dummy" and attempted a takeoff — and quickly found out why the name came into use. The bad air cartwheeled me down a steep slope about 30′, but somehow I was able to stop there before reaching the really steep part. It was a struggle to get my undamaged kite back to the top, but no struggle to put off any more flying attempts for that day.

Enthusiasm was running high and Helen and I decided to make our next vacation a hang gliding vacation. We loaded up our newly acquired high performance kite on the Van and headed south to fly California's better known gliding sites of the time, Fort Funston, Big Sur, Santa Barbara, Torry Pines, etc.

The Torrey Pines cliffs overlook Black's Beach, which is a well known nudist beach. The cliffs are high enough that the nudists are pretty well beyond exciting anyone's prurient interests from the top. I'm not quite sure why, but something caused me to end my first flight with a landing down on the Beach. Gracious! I had to disassemble the kite and walk back through all those nudies. The flying was simply great, 20 to 45 minute flights were not at all unusual. It was a memorable two weeks.

Flying the hang gliders was an experience unlike any other. In the beginning I must admit to considerable concern. Standing at the edge of a precipice and looking down several hundred feet to the bottom could give pause. It only took a few jumps to allay the doubts, however. Indeed, complacency could become an enemy. I recall reaching the point where the pleasure of feeling the lift of the air on the cloth wings, and the joy of the whisper quiet flight pushed all concern out of my mind. On one takeoff, I was in such a state of euphoria that I failed to give proper attention to the need for adequate preparation. The launch into flight was so poorly done that I momentarily dropped like a rock and once more bounced my butt off that ledge about 40 feet below. It was a very painful touch and go. The discomfort of a sore butt for the remainder of the way down made for an indelible reminder that too much bliss could be fatal.

Shifting the weight of your body furnishes more than enough control over the hang glider in calm air. There can be times that a spot of rough air can make it a little less than adequate, however.

I recall one flight off the Sand Dollar site that presented such a moment. The area is such that you have to fly straight out for a fair distance before you can even see the landing spot. As I neared the end of that straight out portion, I must have hit a small dust devil (with no dust). The hang glider quickly went into some fairly steep maneuvers that were well beyond what I would have liked, or intended. My thought at that moment was that if Helen was still watching, she probably thought I was showing off with a little aerobatic demonstration.

By the time she arrived to pick me up I had all but forgotten about it. But sure enough, her first comment was: "I was sure surprised to see you doing all that aerobatic stuff." I hastened to reply that I was too! Another time at Fort Funston, conditions were perfect for soaring. The breeze was steady and smooth, and I was really enjoying the flight. But, for no reason that I could determine, a weird spot of air just suddenly took over and landed me on top of the cliff, whether I was ready or not. It must have looked like I had intended to do just that to the crowd so I merely stepped off the cliff again and continued the flight.

In spite of those moments that made you wonder about the wisdom of leaping off cliffs, the fun parts like the time Lan and I flew formation down to the landing site at Usal more than made up for them. Hang gliding had been a lot of fun indeed.

Getting to our flying site was quite a chore and the weather was always unpredictable. So, with the growing sophistication of powered hang gliders, our interest began turning in that direction. Where ultralights — as the powered hang gliders came to be known — lacked the quiet, peaceful flight of hang gliding, they did offer the ability to fly off our nearby airport where weather was more easily evaluated, and still offered that open, free feeling that's not found with conventional light aircraft. It seemed like a good thing to try.

The final impetus for a switch in allegiance came with Al Trefethen's acquisition of a new Quicksilver MX. Somehow I found myself, a totally green would-be ultralight pilot, ready to test a newly assembled ultralight. At least I was pretty sure it would fly, although I would have to approach it with caution. This model did not have standard airplane controls. The rudder pedals operated wing spoilers instead of the rudder, and the rudder was controlled by the stick action that normally operated the ailerons.

It proved to be no problem on such a docile craft and I was hooked, we had to have one. This desire lead to what has to be described as the saga of the "Unlucky Ultralight." We acquired the remains of a brand new $5,000 Quicksilver from a fellow who thought he could fly the thing without proper instruction. It only took about five seconds of his first effort to provide the knowledge that he couldn't. It also provided him with a broken back.

Helen in our modified MX ultralite.

The wings and tail were salvageable, and the engine was still brand new, so it made a cheap and easy way to design my own fuselage that would incorporate the changes that I felt were very desirable. What I wanted was a steerable nosewheel, standard aircraft controls, and a method for easier break-down and assembly. We planned on keeping it at home under cover and trailering it to the airport when we wanted to fly.

After two weeks of designing and building, it was time to test fly my creation. Whenever these test flights arise there's always that little twinge of doubt — do I really want to do this? Of course! The feeling of self satisfaction and accomplishment when that first flight is over is always worth the concern. At least this test flight was bound to go at a much slower pace than most of my previous ones.

The flight went off without a hitch and the only fix needed was a slight adjustment to the elevator trim tab. We now had a good flying ultralight and we all were having a ball. Helen, Lan and I seemed to fly it more than the Piper Vagabond that we had at the time.

We enjoyed the ultralight a lot but I could see even more room for improvement in the area of portability. I was more than willing to start on another one, so when the chance came to trade this one off to Terry Parsons for a lot of good airplane parts, I took it. Terry and his wife Marilyn lived in Canada and we had been friends a long time. He was eager and kind enough to volunteer for the task of hauling his stuff down and trailering the ultralight back to Canada. Unfortunately, on the way home he made the mistake of pulling into a motel that had a carport-like roof in front of the office that proved to be a mite too low for the ultralight. The ensuing collision completely destroyed the fuselage.

Undaunted, and being at least my equal in the design and building department, he undertook the task of designing and building an entirely different fuselage after finally getting the thing home. Since Terry tends to work at a much more leisurely pace than I do, it took him quite a while to complete the project. Just by chance, Helen and I happened to head up that way just as he was getting ready to take it down to the local field.

The Canadian laws wouldn't allow him to fly it until he had finished his dual instruction in ultralights so he prevailed upon me to do the test flying chores. I wasn't cleared to fly in

Canada either, but in the interest of having a tested ultralight for Terry to fly on one of his very early solo flights, I agreed. It was a good decision.

After a couple of short lift offs, I went for it all, only to have the engine quit about 50′ in the air! Fortunately, there was another field just ahead and no problem getting it down. Gads, at 65 I could do with some other form of aerobic heart stimulation. After removing an offending piece of flapping velcro from the fuel vent and adjusting the horizontal stabilizer, I was able to make a few more fault free flights and pronounced it O.K. for him to fly after his dual instruction.

A few weeks after we returned home, we received a letter from Terry with a picture of the badly damaged ultralight! It seems that on his very first solo flight in his newly modified ship, something went wrong and he landed in a flooded field. The reasons for this are still not quite clear in my mind. At any rate, the only way he could retrieve the undamaged ship was to have it hauled out by helicopter. Apparently no one anticipated the damage the helicopter's down wash could do, but it did plenty.

I wrote back and suggested that he should abandon the unlucky ultralight and just buy a brand new one. I don't know whether my suggestion was the key or not, but he did just that and said we were welcome to the remains of the old one if we would just come up and get them. Since we had had very good luck with it, we decided to take him up on the offer. It gave me the chance to rebuild and incorporate all of my ideas for a more portable ultralight.

It was finished in a few weeks and once again I had to dig up the fortitude for another first test flight. It started out so beautifully that I swung out a little farther than usual to enjoy the view of the beach instead of following my usual habit of keeping my first flights within easy reach of the airport.

That's when it became my turn for the Unlucky Ultralight to bite the hand that built it. The drive belt on the reduction system broke and made the thing an instant glider — with no place to land. I tried to stretch the glide to an open spot but there were just too many things in the way. The resultant impact was very hard on the ultralight, and not too easy on one or two of this senior citizen's internal organs. We decided it was time for us to get rid of the damaged mess (I kept the organs) and move on to something else. I was only sore for a couple of days and the ultralight would have been repairable — but I think the skies have seen the last of the "Unlucky Ultralight."

June 1, 1987 marked the 50th anniversary of the beginning of my flying career. From that tentative start it would have been impossible to even imagine the opportunities aviation had to offer me in the years to come. It's been most gratifying to have flown such a variety of airplanes from a replica of a Glenn Curtis pusher to the fighter planes of World War II.

It's impossible to say just what phase of flying I favored the most. I suppose the answer would have to be that it was just whatever I was involved with at the moment.

Typesetting by T.A. Douglass, 14722 LeClerc Rd. Newport WA 99156
All typesetting was done in WordPerfect and printed on an HP LaserJet III printer with a PacificPage Postscript cartridge.
The type is Palatino for the body text and Palatino Italic for the captions